C++语言与机器视觉编程实战

汪兆栋　黄学雨　陈　锐
黄小燕　李永高　万君社　编著

北京邮电大学出版社
www.buptpress.com

内 容 简 介

本书以 C++ 为编程语言，以 QT 为软件开发框架，采用 OpenCV 图像算法和 Tensorflow 深度学习开发平台，主要内容包括 C++ 开发环境的搭建、QT 开发相机驱动、光源控制、通信控制、用 OpenCV 进行定位和测量、用 Tensorflow 进行图像分类和分割。本书可以作为高等院校机器人专业、自动化控制专业、电器控制专业、智能制造专业、大数据专业等专业教材及相关职业的培训教材，也可以作为其他专业选修或自学教材。

图书在版编目(CIP)数据

C++语言与机器视觉编程实战／汪兆栋等编著．－－北京：北京邮电大学出版社，2024.1（2024.7重印）
ISBN 978-7-5635-7162-8

Ⅰ．①C… Ⅱ．①汪… Ⅲ．①计算机视觉－C++语言－程序设计 Ⅳ．①TP302.7②TP312.8

中国国家版本馆 CIP 数据核字(2023)第 249144 号

策划编辑：姚 顺 刘纳新 责任编辑：廖 娟 责任校对：张会良 封面设计：七星博纳	
出版发行：北京邮电大学出版社	
社　　址：北京市海淀区西土城路 10 号	
邮政编码：100876	
发 行 部：电话：010-62282185　传真：010-62283578	
E-mail：publish@bupt.edu.cn	
经　　销：各地新华书店	
印　　刷：河北虎彩印刷有限公司	
开　　本：787 mm×1 092 mm　1/16	
印　　张：16	
字　　数：422 千字	
版　　次：2024 年 1 月第 1 版	
印　　次：2024 年 7 月第 2 次印刷	

ISBN 978-7-5635-7162-8　　　　　　　　　　　　　　　　　定 价：59.00 元

・如有印装质量问题，请与北京邮电大学出版社发行部联系・

前　　言

随着人工智能和智能制造的发展，人工智能与机器视觉的结合应用是本书介绍的重点。通过深度学习技术，我们可以将机器视觉中的区域模型、缺陷模型和特征模型，通过神经网络训练后，进行机器视觉的定位、测量、识别和检测。深度学习视觉与传统的机器视觉相比，开发周期缩短了80%，对人员技术要求降低了80%，而执行效率在GPU的加速条件下提高了80%。

本书共分为20章，建议每章学习3～4学时，共计60～80学时。本书的学习总体上可以分为以下五部分。

第1～3章是对编程语言能力的培养。该部分详细讲解了编程环境的搭建、C语言的基础语法和C++语言的特点。

第4～5章是对软件开发能力的培养。该部分主要介绍了代码管理的方法、跨平台QT软件的学习，特别是多线程软件的构建方法。

第6～11章是对机器视觉和图像开发能力的培养。基于OpenCV图像库技术，图像的采集、定位、测量和字符识别是该部分的重点。

第12～19章主要介绍了人工智能技术在视觉检测中的应用。特别是以ChatGPT为代表的深度学习技术极大地推动了人工智能技术在工业检测中的应用。

第20章详细讲解了机器视觉方案设计。根据客户的需求，对机器视觉项目的需求进行拆解后，选择合适的相机、光源、镜头，最后形成机器视觉方案。

感谢广东海光云科技股份有限公司对本书出版的大力支持，特别感谢华南理工大学游林儒教授的指导。由于编者水平有限，书中难免存在不足和错误，恳请读者批评、指正。

作　者

目　　录

第 1 章　C++编程环境的搭建 ··· 1
1.1　Win 10 环境下 Visual Studio 2019 安装 ································ 1
1.2　Win 10 环境下 Qt 5.15.2 安装 ·· 2
1.3　第一个程序 hello world! ··· 7
思考题 ·· 11

第 2 章　C 语言 ·· 12
2.1　C 语言的第一个简单实例 ·· 12
2.2　C 语言数据类型 ··· 13
2.3　输入与输出 ··· 14
2.4　控制结构 ·· 17
2.5　函数 ·· 23
2.6　数组 ·· 25
2.7　地址与指针 ··· 27
2.8　结构体和共用体 ··· 29
思考题 ·· 33

第 3 章　C++语言 ··· 36
3.1　类和对象 ·· 36
3.2　类的继承 ·· 37
3.3　类的多态 ·· 38
3.4　类的模板 ·· 38
思考题 ·· 42

第 4 章　CMake 构建系统 ·· 44
4.1　用 CMake 构建一个 HelloWorld ··· 44
4.2　CMake 与 OpenCV ··· 47
4.3　CMake 与 QT ·· 50
4.4　CMake 与 Tensorflow ·· 53
思考题 ·· 60

第 5 章　QThread 多线程 ·· 61

5.1　基于 QThread 多线程 ·· 61
5.2　基于 QMutex 和 QMutexLocker 多线程 ·· 67
5.3　基于 Ontimer 显示多线程 ··· 73
5.4　基于 Producer 与 Consumer 显示多线程 ·· 79
思考题 ·· 86

第 6 章　QT、OpenCV、HIK 图像采集系统 ·· 87

6.1　基于 Hik-QT-IMG 软触发采集图像 ··· 87
6.2　纯虚函数 Virtural 接口方法 ·· 89
6.3　环形分区光源合成算法 ··· 104
思考题 ·· 107

第 7 章　C++和 OpenCV 视觉九点标定和五点旋转标定 ······························ 108

7.1　背景介绍 ·· 108
7.2　视觉建模 ·· 108
7.3　旋转标定 ·· 111
7.4　代码分析 ·· 113
思考题 ·· 115

第 8 章　C++和 OpenCV 实现卷绕视觉纠偏 ··· 116

8.1　纠偏需求分析 ·· 116
8.2　图像处理流程 ·· 118
8.3　左边基准线 ·· 119
8.4　右边基准线 ·· 122
8.5　中间特征线 ·· 125
思考题 ·· 129

第 9 章　C++和 OpenCV 实现键盘缺陷检测 ··· 130

9.1　需求分析 ·· 130
9.2　图像 ROI ·· 130
9.3　分区建模 ·· 131
9.4　模板 XML ·· 132
9.5　缺陷算法 ·· 134
思考题 ·· 136

第 10 章　SIFT 和 ANN 实现任意顺序图像的合成 ·· 137

10.1　算法原理 ·· 137

10.2	特征寻找	139
10.3	三张合成	150
10.4	五张合成	151
10.5	七张合成	153
思考题		154

第 11 章　采用 Tesseract、BP、DP 进行 OCR 识别 ... 155

11.1	字符识别方案选择	155
11.2	采用 Tesseract 识别英文和数字	155
11.3	采用 BackPropagation 识别喷码字符	156
11.4	采用 DeepLearning 识别中文手写体	158
思考题		163

第 12 章　Python 和 Tensorflow 深度学习分类网络 ... 164

12.1	分类网络	164
12.2	模型网络	164
12.3	推理执行	165
12.4	实验报告	166
思考题		173

第 13 章　Python 和 Tensorflow 深度学习分割网络 ... 174

13.1	图像标注	174
13.2	图像提取	177
13.3	模型训练	178
13.4	版本控制	179
13.5	安装过程	180
思考题		187

第 14 章　C#和 C++接口设计 CLR ... 188

14.1	接口 CLR	188
14.2	图像转换	189
14.3	字符串转换	191
14.4	框架设计	191
思考题		192

第 15 章　基于 C#深度学习焊点缺陷检测 ... 193

15.1	C#深度学习开发流程	193
15.2	焊点检测任务	193

15.3 深度学习焊点检测分析 ··· 195
思考题 ··· 196

第 16 章 基于 Pyqt5 和 OpenCV 图像处理平台 ··································· 197

16.1 说在前面 ··· 197
16.2 编写目的 ··· 197
16.3 产品范围 ··· 197
16.4 阅读建议 ··· 197
16.5 运行环境、库、GPU 要求 ··· 197
16.6 环境配置 ··· 198
16.7 框架介绍 ··· 201
16.8 未来计划 ··· 201
思考题 ··· 201

第 17 章 基于 C++视觉软件系统构建 ··· 202

17.1 基于 QThread 的日志跟踪 ··· 202
17.2 基于海康工业相机的图像采集 ·· 204
17.3 基于 Basler 工业相机的图像采集 ··· 205
17.4 基于串口通信和光源控制系统 ·· 207
思考题 ··· 209

第 18 章 深度学习用于焊缝检测实验报告 ··· 210

18.1 检测条件 ··· 210
18.2 测试结果 ··· 210
18.3 存在问题 ··· 210
18.4 改善方法 ··· 218
18.5 实验结论 ··· 218
思考题 ··· 218

第 19 章 Win 10 环境下深度学习训练环境构建 ·································· 219

19.1 Python ··· 219
19.2 labelme ·· 222
19.3 Numpy ··· 224
19.4 Microsoft Visual Studio 2019 ·· 225
19.5 Microsoft Visual Studio Code ·· 227
19.6 TensorFlow ··· 228
19.7 OpenCV ··· 229
19.8 Cuda ··· 230

19.9	Cudnn	234
19.10	Keras	235
思考题		236

第 20 章　视觉方案 ································ 237

20.1	项目需求	237
20.2	打光图片	238
20.3	结论分析	240
20.4	硬件配置	240
20.5	硬件安装	241
20.6	安装图纸	241
20.7	专业术语	243

第 1 章
C++编程环境的搭建

本章内容是基于 Win 10 环境下，Visual Studio 2019 和 Qt5.15.2 的安装说明，最后用一个 hello world! 的实例来验证编程环境。

1.1 Win 10 环境下 Visual Studio 2019 安装

Visual Studio(简称 VS)是美国微软公司的开发工具包系列产品。VS 是一个基本完整的开发工具集，它包括了整个软件生命周期中所需要的大部分工具，如 UML 工具、代码管控工具、集成开发环境(IDE)等。

1. 进入 Microsoft 下载官网，网址链接：https://visualstudio.microsoft.com/zh-hans/vs/older-downloads/。

2. 选择需要安装的版本并下载。Visual Studio 2019 分为社区版、专业版和企业版，其中社区版可免费使用，此处安装的是社区版。

3. 点击运行安装程序，进入安装。出现图 1-1 所示的界面后，点击继续。

图 1-1 进入准备安装的画面

提取文件并进入正常安装后，选择个人所需要的开发负载、组件、语言包(此处选择"使用 C++的桌面开发"和"通用 Windows 平台开发")，选择安装位置(默认 C 盘)，并选择下载时安装。开始安装时，可以选择安装的组件。如图 1-2 所示。

4. 重启。安装完成后提示重启，重启计算机后，安装就完成啦！

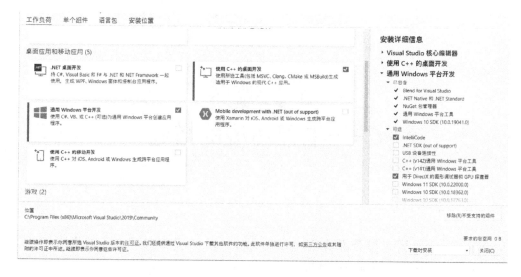

图 1-2　选择 C++需要的组件

5. 完成。具体的安装过程，可以参考原文链接：https://blog.csdn.net/weixin_42826790/article/details/109197287。

1.2　Win 10 环境下 Qt 5.15.2 安装

第一步：开启 Win 10 计算机后，打开 Qt 在线安装程序下载网站：Index of /official_releases(qt.io)。如图 1-3 所示。

Index of /qtproject/official_releases/

../		
QtForPython/	12-Apr-2022 20:00	-
additional_libraries/	03-Mar-2021 16:18	-
gdb/	17-Nov-2014 19:42	-
jom/	12-Dec-2018 21:13	-
online_installers/	13-Mar-2023 17:43	-
pyside/	30-Nov-2015 19:39	-
qbs/	03-Apr-2023 19:11	-
qt/	03-Apr-2023 12:37	-
qt-installer-framework/	13-Mar-2023 15:55	-
qt3dstudio/	28-Oct-2020 20:22	-
qtchooser/	08-Oct-2018 12:53	-
qtcreator/	28-Mar-2023 16:22	-
qtdesignstudio/	14-Jul-2022 01:21	-
vsaddin/	24-Feb-2023 17:51	-
timestamp.txt	06-Apr-2023 01:00	11

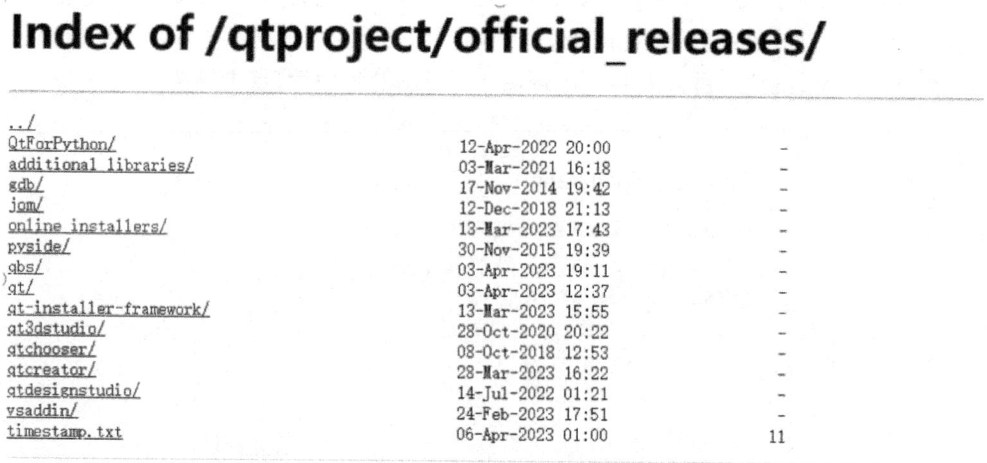

图 1-3　Qt 在线安装程序选择界面

选择 online_installers 进入在线程序，里面有三个版本，分别为 Linux 版本、Mac 版本和 Windows 版本，可根据需要下载(本次下载的是 Windows 版本)。如图 1-4 所示。

打开安装程序，登录 Qt 账户(若无 Qt 账户，则自行注册)，点击“下一步”。如图 1-5 所示。

勾选以下两个选项，点击“下一步”。如图 1-6 所示。

Index of /qt/official_releases/online_installers/

../
qt-unified-linux-x64-online.run   13-Mar-2023 16:04   57561126
qt-unified-mac-x64-online.dmg   13-Mar-2023 16:04   19289770
qt-unified-windows-x64-online.exe   13-Mar-2023 16:04   43016864

图 1-4 Windows 版本在线安装程序选择界面

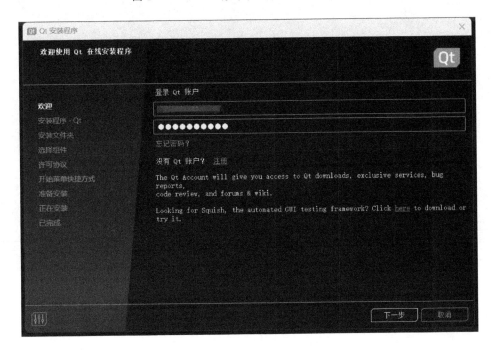

图 1-5 Qt 安装程序登录界面

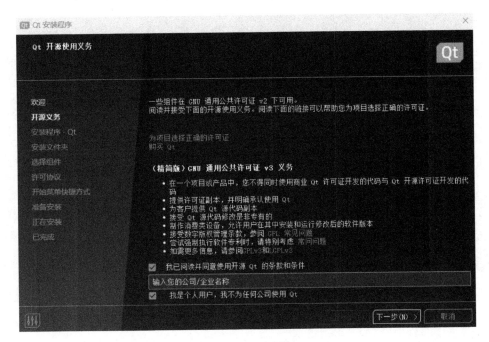

图 1-6 Qt 开源义务

继续点击"下一步"。如图1-7所示。

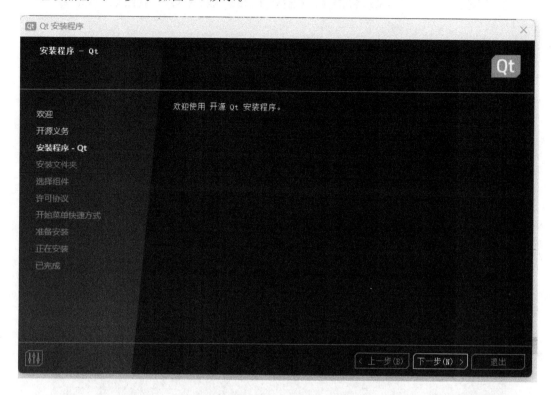

图1-7 Qt安装程序

安装软件将会自动提取远程数据,完成后将弹出图1-8所示的界面,勾选后点击"下一步"。

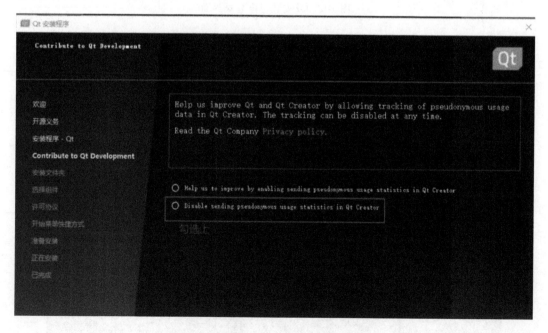

图1-8 Qt安装程序

接着选择 Qt 的安装目录。如图 1-9 所示。

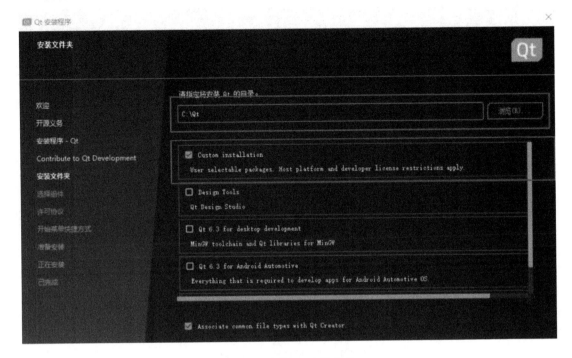

图 1-9　选择 Qt 安装目录

选择需要安装的组件。如图 1-10 和图 1-11 所示。

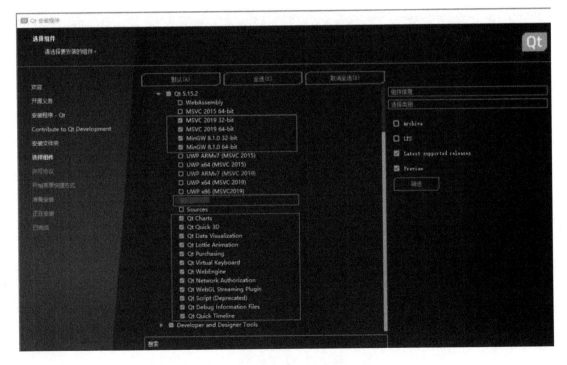

图 1-10　Qt 组件选择安装界面

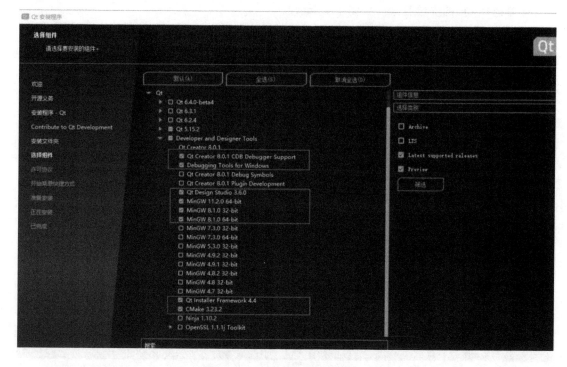

图 1-11 Qt 组件选择安装界面

选择完组件后,同意协议,准备安装。如图 1-12 所示。

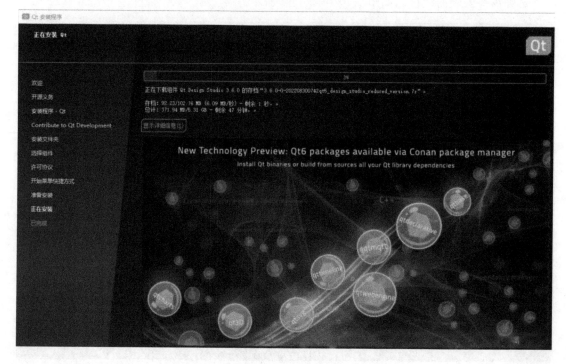

图 1-12 Qt 安装界面

安装完成后，最后一步就是设置 Qt 环境变量。右击此计算机，选择"属性"，然后单击"高级系统设置"，选择"环境变量"，双击系统变量中的 path，新建一个环境变量，并输入 Qt 的 bin 文件的路径。如图 1-13 所示。

图 1-13　编辑 Qt 环境变量

1.3　第一个程序 hello world！

第一步：打开 Win 10 计算机。
第二步：双击打开 Visual Studio 2019。如图 1-14 所示。

图 1-14　Visual Studio 2019 桌面图标

第三步：点击"继续"，但无须代码，进入编程界面。如图 1-15 和图 1-16 所示。
第四步：开始一个"Helloworld"程序如图 1-17 所示。
① 首先，输入项目名称，即输入 Helloworld。如图 1-18 所示。
② 然后，点击"创建"，出现以下界面，如图 1-19 所示。

图 1-15 Visual Studio 2019 打开界面

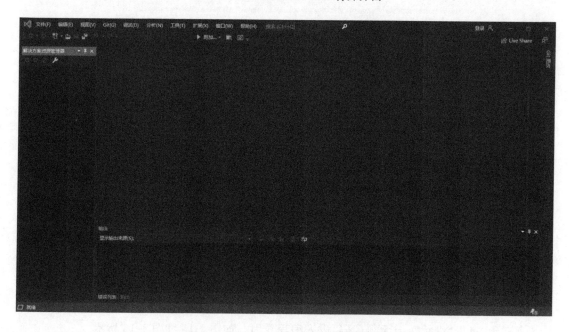

图 1-16 Visual Studio 2019 编程界面

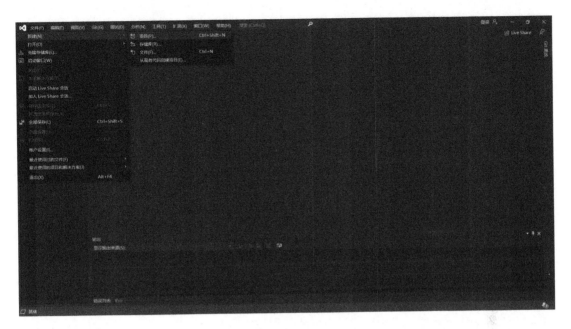

图 1-17　新建项目

图 1-18　配置新项目

C++语言与机器视觉编程实战

图 1-19 Helloworld 项目新建完成

③ 运行以下代码,得出结果。如图 1-20 所示。

```
int main()
{
    std :: cout <<"Hello World! \n";
}
```

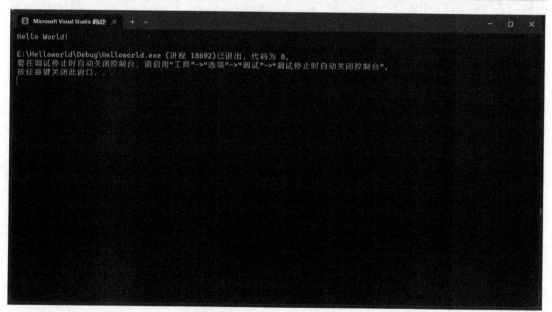

图 1-20 Helloworld 项目输出结果

祝贺你开启了 C++第一个完整程序,hello world!

思 考 题

简答题
1. C++的版本是多少？它们之间的区别有哪些？通常用哪个版本？
2. VS 和 QT 的关系是什么？
3. 环境变量配置的方法和目的是什么？

第 2 章 C 语 言

C语言是一门通用的计算机汇编语言,在系统软件与应用软件中有着广泛的应用。C语言具有灵活方便,数据处理能力强,数据类型丰富,适用范围广等特点,它既具备高级语言的特征,又具有汇编语言的特点。使用C语言编写程序对于我们以后更深层次的学习和工作都有很大的帮助,让我们一起来学习C语言的相关知识吧!

2.1 C语言的第一个简单实例

到底C语言是什么样的呢?让我们来看一个简单的C语言程序吧!

```c
#include<stdio.h>           /*引入头文件*/
int main()                  /*一个简单的C程序*/
{
    int number;             /*定义个名字叫做 number 的变量*/
    number = 2023;          /*给 number 赋一个值*/
    printf("This year is %d\n",number);
    return 0;
}
```

输出结果如图 2-1 所示。

图 2-1 C语言程序输出结果

#include<stdio.h>是这个程序的第一行,其主要作用是引入 stdio.h 文件的全部内容,方便后面函数的调用。在其他程序中,也可以引入其他的头文件来代替 stdio.h,调用其他头文件内容。

int main 在程序中声明了一个主函数 main,一个C语言程序总是从 main 函数开始的。int 指明了 main 函数返回给操作系统的返回值的数据类型,必须把结果返还给系统,系统才能判断输出的结果是否正确。这里的 int 也可以用 float 等其他数据类型来代替。

{…}在 C 语言中表示函数体的开始与结束,且只有花括号能起到这种作用,小括号和中括号都没有这种作用。

int number;是程序中的声明语句,其主要作用是在函数中命名一个 number 的变量,并确定其数据类型为整型数据。在 C 语言中,所有的变量在使用之前都必须进行定义,并确定其数据类型。

number = 2023;是程序中的赋值语句,其主要作用是把 2023 赋值给变量 number。

printf("This year is %d\n",number);是程序中的输出函数,主要是将括号中的内容输出到屏幕上。%d 是一个占位符,作用是指出 number 的位置,且输出结果被 2023 代替。\n 是一个换行符,其作用为输出结果完毕之后开始新的一行。

return 0;是一个返回语句,表示程序的正常退出。

/*…*/为程序的注释部分,其主要作用是解释程序中代码的作用,方便后续的修改和使用。

通过对这个简单例子的分析,相信大家对 C 语言有了一个初步的理解。

2.2 C语言数据类型

在 C 语言中,数据的两种表现形式是常量和变量。在整个程序运行过程中,值没有发生改变且不能改变的称为常量,值可以改变并且有可能改变的称为变量。

在 32 位操作系统中,常见编译器的数据类型大小及表示的数据范围如表 2-1 所示。

表 2-1 C语言的数据类型

类型名称	类型关键字	占字节数	其他叫法	表示的数据范围
字符型	char	1	signed char	$-128 \sim 127$
无符号字符型	unsigned char	1	none	$0 \sim 255$
整型	int	4	signed int	$-2\,147\,483\,648 \sim 2\,147\,483\,647$
无符号整型	unsigned int	4	unsigned	$0 \sim 4\,294\,967\,295$
短整型	short	2	short int	$-32\,768 \sim 32\,767$
无符号短整型	unsigned short	2	unsigned short int	$0 \sim 65\,535$
长整型	long	4	long int	$-2\,147\,483\,648 \sim 2\,147\,483\,647$
无符号长整型	unsigned long	4	unsigned long	$0 \sim 4\,294\,967\,295$
单精度浮点数	float	4	none	$-3.4E38 \sim 3.4E38$ (7 digits)
双精度浮点数	double	8	none	$-1.79E+308 \sim +1.79E+308$ (15 digits)
长双精度浮点数	long double	10	none	$-1.2E+4932 \sim +1.2E+4932$ (19 digits)

认识了上面的数据类型之后,我们就可以根据不同程序的需求选择不同的数据类型来使用。下面让我们通过一个简单的例子来学习一下如何使用数据类型吧!

```
#include<stdio.h>
int main()
{
    int a = 200, b = 300;
    int c;
    c = a + b;
    printf("a+b=%d\n", c);
    return 0;
}
```

输出结果如图 2-2 所示。

图 2-2　a＋b 的输出结果

在上面例子中,我们定义了 a,b,c 三个整型变量,以实现 c=a+b 的功能,并且通过 printf 函数输出结果。如果我们需要使用超过 int 数据范围的数据时,只要根据表 2-1 选择适合我们使用的数据类型即可。

在 C 语言中,除了数据类型外,我们在定义变量、表达语句功能和对一些文件进行预处理时还需要用到一些关键字,但我们自己定义的变量函数名等不可以与关键字同名。

C 语言中的 32 个关键字如表 2-2 所示。

表 2-2　C 语言中的 32 个关键字

auto	double	int	struct
break	else	long	switch
case	enum	register	typedef
char	extern	return	union
const	float	short	unsigned
continue	for	signed	void
default	goto	sizeof	volatile
do	if	static	while

2.3　输入与输出

在 C 语言程序中,我们经常需要对数据进行输入和输出操作,这个时候就要使用特定的函数对数据进行处理。如果只需要处理单个字符,我们可以使用 getchar 函数和 putchar 函数。如果需要按照指定的格式处理数据,我们可以使用 printf 函数和 scanf 函数。

getchar 函数是字符输入函数,其作用是接收用户输入的一个字符,输入的格式为:a=getchar()。getchar 会以 ASCII 码的形式返回接收到的字符。

putchar 函数是字符输出函数,其作用是向终端输出一个字符,输出的格式为:putchar(a)。当 a 为一个被单引号引起的字符时,输出结果为该字符;当 a 为一个介于 0~127 之间的十进制整数时,输出结果为对应的 ASCII 代码。

下面我们通过一个例子来学习一下 getchar 函数和 putchar 函数的使用。

```c
#include<stdio.h>
void main()
{
    char a,b,c,d;
    printf("请输入字符:\n");
    a = getchar();
    b = getchar();
    c = getchar();
    d = getchar();
    putchar(a);
    putchar(b);
    putchar(c);
    putchar(d);
}
```

输出结果如图 2-3 所示。

图 2-3　getchar 函数和 putchar 函数实例

这个程序实现了单个字符的输入和输出,当我们运行程序之后,通过 getchar 函数获取我们输入的字符,然后通过 putchar 函数输出到屏幕上。我们也可以输出其他的字符,但是需要注意的是,一个函数只能对应一个字符。

除了单个字符的输出,我们还可以使用 printf 函数和 scanf 函数进行特定的格式输出。

printf 函数是格式化输出函数,其作用是按照用户设定的格式输出数据,输出格式为:printf("格式控制字符串",输出项);格式控制字符串为格式字符串和非格式字符串两种。格式字符串是以%开头的字符串,根据在%后面所带的字符串的不同,所输出的数据也不相同。非格式字符串在输出结果时依照原样打印出来。

C 语言中常用的输出格式如表 2-3 所示。

表 2-3　C 语言中常用的输出格式

格式字符	输出格式说明
d,i	以十进制形式输出有符号整数(正数不输出符号)
O	以八进制形式输出无符号整数(不输出前缀 0)
x	以十六进制形式输出无符号整数(不输出前缀 0x)
U	以十进制形式输出无符号整数
f	以小数形式输出单、双精度类型实数
e	以指数形式输出单、双精度实数
g	以%f 或%e 中较短输出宽度的一种格式输出单、双精度实数
C	输出单个字符
S	输出字符串

scanf 函数是格式输入函数,其作用是将数据按格式字符串的格式输入到指定的变量中,输出的格式为:sacnf("格式控制字符串",输入项);scanf 函数在使用时不能显示非格式字符串,输入项中变量的地址由地址运算符和后面的变量名组成。

C 语言中常用的输入格式如表 2-4 所示。

表 2-4　C 语言中常用的输入格式

转换说明符	输出格式说明
%c	把输入解释成一个字符
%d	把输入解释成一个有符号十进制整数
%e,%f,%g,%a	把输入解释成一个浮点数(%a 是 C99 的标准)
%E,%F,%G,%A	把输入解释成一个浮点数(%A 是 C99 的标准)
%i	把输入解释成一个有符号十进制整数
%o	把输入解释成一个有符号的八进制整数
%p	把输入解释成一个指针(一个地址)
%s	把输入解释成一个字符串:输入的内容以第一个非空白字符作为开始,并且包含直到下一个空白字符的全部字符
%u	把输入解释成一个无符号十进制整数
%x,%X	把输入解释成一个有符号十六进制整数

下面我们通过一个例子来学习一下 printf 函数和 scanf 函数。

```
#include<stdio.h>
int main(void)
{
    int ages;
    printf("请输入您的年龄:");
    scanf("%d", &ages);
```

```
        printf("天数为:%d\n", ages * 365);
        return 0;
}
```

输出结果如图2-4所示。

图2-4　printf函数和scanf函数实例

上述程序实现了一个年龄对应天数的计算,通过scanf函数获取输入的年龄,然后通过printf函数输出年龄*365所对应的值,我们可以通过输入不同的年龄值来获取不同的天数。

2.4　控制结构

在C语言程序中,我们经常需要根据不同的条件选择执行不同的代码,这个时候就需要使用控制结构;如果需要根据给定的条件来决定执行哪一段代码,则需要使用选择结构;如果要反复执行某一段程序,则需要使用循环结构。选择结构中最常使用的就是if选择结构和switch选择结构。

if选择结构用于判断给定的条件是否成立,并根据判断结果来控制程序的流程。选择结构有单选择、双选择和三选择三种形式。

三种选择形式的表达方式如下。

(1) 单选择

```
if(表达式) /*若条件成立则执行后面的语句,反之则不执行*/
{
    //语句
}
```

(2) 双选择

```
if(表达式) /*若表达式成立则执行语句1,反之则执行语句2*/
{
    //语句1
}
else
{
    //语句2
}
```

(3) 三选择

```
if(表达式) /* 如果表达式成立则执行语句 1,反之则继续判断表达式 2 是否成立 */
{
    //语句 1
}
else if(表达式 2) /* 如果表达式 2 成立则执行语句 2,反之则继续判断表达式 3 */
{
    //语句 2
}
else if(表达式 3) /* 如果表达式 3 成立则执行语句 3,反之则继续判断下一个表达式 */
{
    //语句 3;
}
else /* 如果以上表达式都不成立 则执行语句 4 */
{
    //语句 4
}
```

下面我们通过一个例子来学习一下 if 选择结构的使用。

```
#include<stdio.h>
int main() {
    int flag = 80;
    if(flag >= 90)
    {
        printf("Good\n");
    }
    else if(flag >= 60)
    {
        printf("passed\n");
    }
    else
    {
        printf("failed\n");
    }
    return 0;
}
```

输出结果如图 2-5 所示。

在上述程序中,由于变量 flag 给定的值为 80,故 flag 的值不符合条件 1 和条件 3,只符合条件 2,因此输出条件 2 的结果 passed。

当分支条件过多时,if 结构会比较复杂,这时我们就需要使用 switch 结构来解决问题。

图 2-5　if 选择结构

switch 结构的一般形式如下。

```
switch(表达式) /*首先计算表达式的值*/
{
    case 常量表达式 1:语句 1;
    case 常量表达式 2:语句 2;
    case 常量表达式 3:语句 3;
    //……
    case 常量表达式 n:语句 n;
    default:语句 n+1;
}
```

下面我们通过一个例子来学习一下 switch 结构的使用。

```
#include<stdio.h>
int main()
{
    int day = 0;
printf("请输入一个数字\n");    //让用户输入一个数字
scanf("%d", &day);
switch(day)
{
case 1:printf("星期一\n");
    break;
case 2:printf("星期二\n");
    break;
case 3:printf("星期三\n");
    break;
case 4:printf("星期四\n");
    break;
case 5:printf("星期五\n");
    break;
case 6:printf("星期六\n");
    break;
case 7:printf("星期天\n");
    break;
}
return 0;
}
```

输出结果如图2-6所示。

图2-6 switch结构

程序运行时,day=0时,仅输出"请输入一个数字"。当输入day=1或者其他数字时,输出结果为"星期一"或者其他。即switch语句是将表达式的值与常量的值进行比较,并输出对应的结果。

break语句通常用于跳出switch结构或循环结构,用于提前结束switch结构或循环。

当需要反复执行某一段程序时,我们就需要使用循环结构。C语言提供了三种循环结构,分别为while循环、do while循环和for循环。

while循环是一种基本的循环结构。当满足条件时进入循环,并且重复执行循环的部分,直到条件不满足时跳出循环。while循环在进行循环之前会判断条件是否成立,如果不成立则不执行循环,因此while循环可能一次也不循环。

while循环的一般形式如下。

```
while(表达式)
{
    循环体语句
}
```

下面我们通过一个例子来学习一下while循环的使用。

```
#include<stdio.h>
int main()
{
    int i=0;
    while(i++<20)
    {
        printf("count %d\n",i);
    }
    return 0;
}
```

输出结果如图2-7所示。

从结果可以看出,当i<20时,while语句会一直循环进行,当i=20时循环结束。

do while循环是在判断表达式为假(或0)之前重复执行循环部分。do while循环与while循环相比,do while循环必须在执行一次循环之后才决定是否要再次执行循环,因此循环至少要被执行一次。

do while循环的一般形式如下。

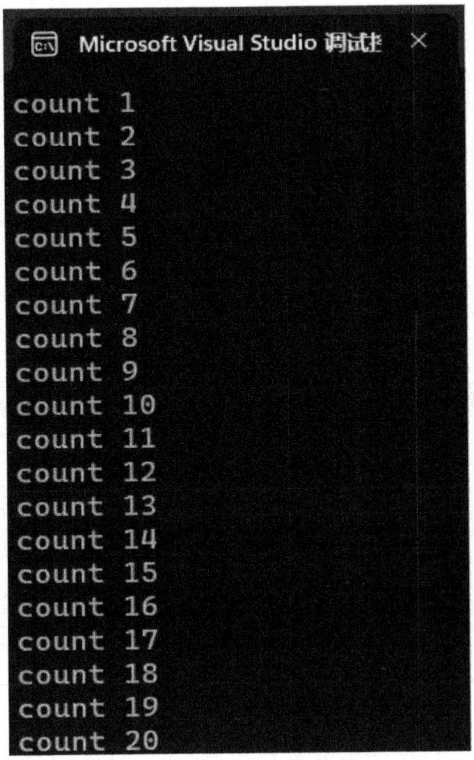

图 2-7 while 循环

```
do
{
    循环体语句
}while(表达式);
```

下面我们通过一个例子来学习一下 do while 循环的使用。

```
#include<stdio.h>
int main()
{
    int i = 0;
    do
    {
        printf("count %d\n",i);
    }while(i++<20);
    return 0;
}
```

输出结果如图 2-8 所示。

与 while 循环相比，do while 循环多了一个 count 0 的输出，是由于 do while 循环必须在执行一次循环之后才决定是否进行下一步循环，因此在第一次执行时输出了 count 0。

for 循环是由三个控制表达式来控制循环过程的。初始化表达式是在循环开始前执行的，如果条件成立则执行一次循环，然后更新表达式并与判断条件比较，条件成立则循环继续，反

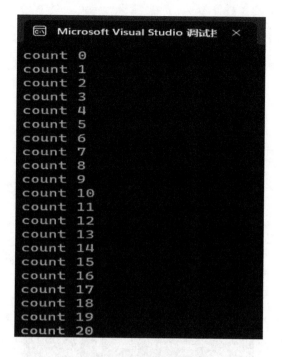

图 2-8　do while 循环

之则循环结束。for 循环在进行循环之前会判断条件是否成立,如果不成立则不执行循环,因此 for 循环可能一次也不循环。

for 循环的一般形式如下。

```
for(初始化表达式;判断表达式;更新表达式)
{
    循环体语句
}
```

下面我们通过一个例子来学习一下 for 循环的使用。

```
#include<stdio.h>
int main()
{
    int i;
    for(i=0;i<20;i++)
    {
        printf("count is %d\n",i);
    }
    return 0;
}
```

输出结果如图 2-9 所示。

由于在执行循环语句前,初始表达式会先执行一次,故当 i=20 时,判断表达式不成立,不再进行循环,即当输出 count is 19 之后,循环结束。

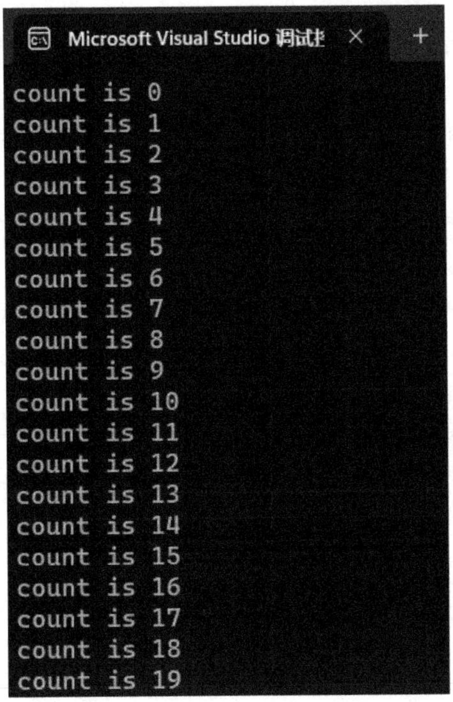

图2-9　for循环

2.5　函　　数

函数是C语言程序的基本模块,程序的许多功能是通过对函数模块的调用来实现的,学会编写和调用函数可以提高编程效率,实现需要的功能。

一个最简单的程序只包含了一个主函数main(),但如果要实现一个具有完整功能的程序则通常是由主函数调用其他函数,多个函数一起组合实现的。

函数的定义的一般格式如下。

```
返回值类型 函数名
{
    说明语句
    执行语句
}
```

返回值类型是指输出结果返回操作系统值的类型。返回值的数据类型通常为int、float、char等。当不指明函数类型时,系统默认的数据类型为整型数据。

函数名是用户自定义的标识符,用于标识函数。在程序中调用函数时需要使用我们定义的函数名。函数名本身也代表了该函数的地址。

花括号包含起来的部分为函数的主体,即函数体。函数体中包含执行语句和说明语句,确定了该函数应完成的运算和执行的动作,体现了函数的功能。

当我们的程序需要实现多个功能时,就需要由主函数来调用其他函数组合实现,这个过程被称为函数调用。

在 C 语言中,通常使用函数表达式、函数语句和函数实参的方式调用函数。

函数表达式:函数以表达式的形式出现在主函数中,通过函数返回值参与表达式的运算,这种方式要求函数有返回值,如 z=max(x,y);,这个函数是一个赋值表达式,把 max 的返回值赋予变量 z。

函数语句:函数语句由所调用的函数的一般形式加分号构成,如 printf("%d",a);和 scanf("%d",&b);,这两种形式都是以函数语句的方式调用函数。

函数实参:这种方式是把该函数的返回值当作实际参数进行调用,因此使用这种方式时需要求被调函数有返回值,如:printf("%d",max(a,b));就是把 max 的返回值作为 printf 函数的实际参数。

在调用函数前,我们应该对被调函数的数据类型进行声明,便于主调函数对这种类型的返回值进行处理。

下面我们通过一个例子来学习一下函数的定义和调用。

```c
#include<stdio.h>
int main()
{
    int a, b;
    int max(int a, int b);
    scanf("%d%d", &a, &b);
    printf("%d", max(a, b));

    return 0;
}
int max(int a, int b)
{
    if(a>b) return a;
    else return b;
}
```

输出结果如图 2-10 所示。

图 2-10 函数的定义和调用

在程序中,我们在主函数 main 中调用了 max 函数,并通过 if 选择结构来选择输出的结果。在程序中,我们使用了函数表达式、函数实参、函数语句三种调用方式来实现比较大小的功能。当我们需要调用其他函数时,也可以参照这个程序对函数进行定义和调用。

在 C 语言程序中,我们在定义和调用函数时经常会用到变量。我们可以通过变量名访问该变量,系统还通过该标识符确定变量在内存中的位置。

在 C 语言中定义了四种存储形式,分别为自动变量(auto)、外部变量(extern)、静态变量(static)和寄存器变量(register),它关系到变量在内存中的存放位置,由此决定了变量的保存时间和变量的作用范围。

我们可以根据变量的保存时间将变量分为静态储存和动态储存。静态存储是指程序运行时为其分配固定的存储空间,动态存储则是在程序运行期间根据需要动态地分配存储空间。

我们也可以根据变量的作用范围将变量分为全局变量和局部变量。全局变量的作用范围为从定义开始到文件的结束,而局部变量则仅在定义的函数或符合语句内有效。

下面我们来了解一下这四种存储属性。

自动变量主要用于储存函数中所有动态局部变量,数据存储态储区中,在函数调用结束后就释放这些存储空间。

外部变量,即全局变量,通常是在文件的开头定义的,有效的作用范围为定义处到文件的末尾。如果要在定义处前引用该变量,则应该在引用之前用关键字 extern 对该变量进行"外部变量声明",表示该变量是已经定义的外部变量。

静态变量主要用于存储在函数结束时有时希望保留原值的变量。在语句块执行期间,static 变量将始终保持它的值,并且初始化操作只在第一次执行时起作用。在随后的运行过程中,变量将保持语句块上一次执行时的值。

寄存器变量是把局部变量指定存放在 CPU 的寄存器中的变量。使用这种方式存储变量可以提高程序的运行速度。但是寄存器变量的储存数量是有限的,不能任意定义存储器变量,且只有局部自动变量和形式参数可以当作寄存器变量。

2.6 数　　组

在 C 语言中,一组同类型的数据的集合称为数组。数组中的每一个数据叫做数组元素,每一个元素可以用数组名和下标表示。数组除了我们常见的一维数组和二维数组之外,还有由多个字符构成的字符数组。

一维数组指的是只有一个下标的数组,它是用来表示一组具有相同类型的数据。

一维数组的一般定义方式为:类型说明符 数组名[常量表达式]。其中,类型说明符的作用是定义数据的数据类型,常量表达式指的是数组中存放的元素的个数。

下面我们通过一个例子来学习一维数组的使用。

```c
#include<stdio.h>
int main()
{
    int a[5]={1,2,3,4,5};
    int i;
    for(i=0;i<5;i++)
        printf("%d\n",a[i]);
```

```
        return 0;
    }
```

输出结果如图 2-11 所示。

图 2-11　一维数组

在这个程序中,我们定义了一个包含五个元素的一维数组,并通过 for 循环对结果进行输出。如果需要修改数组长度,则需要修改 int a[ ]括号中的值,即可修改数组的长度,同时也需要 for 循环的中 i＜n 的条件。

二维数组与一维数组相似,但在用法上比一维数组复杂一点,不过在处理某些问题(如矩阵时),使用二维数组会更加的方便高效。

二维数组的一般定义方式为:类型说明符 数组名[行数][列数]。

下面我们通过一个例子来学习一下二维数组的使用。

```cpp
#include<stdlib.h>
int main() {
    int a[2][3] = {
     {1,2,3},
     {4,5,6}
    };
    for(int i = 0; i < 2; i++) {
        for(int j = 0; j < 3; j++) {
            printf("%d", a[i][j]);
        }
        printf("\n");
    }
    printf("\n");
    return 0;
}
```

输出结果如图 2-12 所示。

图 2-12　二维数组

在这个程序中,输出了一个两行三列的二维数组。通过与一维数组的使用方法对比,我们可以发现两者的使用方法差不多,不过在使用二维数组时多了一个 for 循环,我们需要注意使用时的格式,以免出现错误。

在 C 语言中,当需要输出多个字符时,我们就可以使用字符数组。字符串的各个字符依次存放在字符数组的元素之中。

字符数组的定义和赋值与一维数组和二维数组的使用方法基本相同。

下面我们通过一个例子来学习一下字符数组的使用方法。

```
include<stdio.h>
char main()
{
 char a[] = "How are you";
 int i;
 for(i = 0;i < 11;i + +){
     printf(" % c",a[i]);
 }
 printf("\n");
 return 0;
}
```

输出结果如图 2-13 所示。

图 2-13　字符数组

我们通过这个程序输出了字符串"How are you"。字符数组的使用方法与一维数组和二维数组的使用方法相同,不过在输入字符串时应注意空格键也算是一个字符。

C 语言的字符数组与一维数组和二维数组相比,能够赋予数字字符串常量,在程序中也是十分重要的数组。

2.7　地址与指针

地址是指内存中每个字节的编号,可以用于表示某一点的一个编号。指针则是指数据在内存中的地址。虽然两者都表示某个地址,但性质却不相同。一个代表常量,另一个则代表变量。地址是固定存在的,而指针则是可以变化的,随着赋值不同,指针也会随之变化,这就是指针变量。

指针变量的一般形式为:

类型说明符 * 变量名;

定义中的 * 与前面的类型说明符共同说明这是一个指针变量。

下面我们通过一个例子来学习一下使用指针。

```
#include<stdio.h>
int main()
{
    int num = 123;
    int *p = &num;
    printf("num Address = 0x%x,num=%d\n", &num, num);
    printf("p = 0x%x,*p = %d\n", p, *p);
    return 0;
}
```

输出结果如图 2-14 所示。

图 2-14 指针

通过这种方式我们就可以获取指针变量所对应的地址的值。

我们除了用指针变量来存放数据的内存地址外,还可以通过指针变量来访问数组中的元素。当我们使用一个指向数组元素的指针变量时,它的定义和赋值方法与指针变量相同。

下面我们通过一个例子来学习一下用指针变量来访问一维数组。

```
#include<stdio.h>
int main()
{
    int i;
    int a[5] = {1,2,3,4,5};
    int *p = a;
    for(i = 0;i<5;i++)
    {
        printf("P Value:%d a Value:%d\n", *(p++), *(a+i));
    }
    printf("\n");
    return 0;
}
```

输出结果如图 2-15 所示。

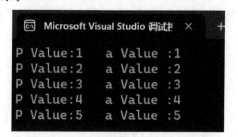

图 2-15 指针变量访问一维数组

程序在运行中用*(p++)或*(a+i)的方式来改变自身值的方式进行移动,从而访问数组中的元素。需要注意的是,数组名的本身值不能改变。

指针变量除了指向数组,也可以指向一个字符串,可以用字符串常量对字符指针进行初始化。

在使用指针变量访问字符串时,应注意字符指针是个变量,我们可以改变指针变量使它指向不同的字符串,但不能改变指针变量所指向的字符串常量的值。

下面让我们来学习一下使用指针变量访问字符串。

```
#include"stdio.h"

void main()
{
char string1[] = "Hello";
char * string2 = "Hello";
printf("%s\n",string1);
printf("%s\n",string2);
}
```

输出结果如图 2-16 所示。

图 2-16 指针变量访问字符串

这个程序使用了两种不同的方式对字符串进行初始化赋值,以及访问字符串的内容。

2.8 结构体和共用体

结构体是由多个数据成员构成的,这些成员可以是不同的数据类型,且可以通过成员名来访问结构体种的元素。

结构体的定义一般形式为:

```
struct 结构类型名
{
    数据类型 成员名1;
    数据类型 成员名2;
    ......
    数据类型 成员名n;
};
```

下面我们通过一个例子来学习一下结构体的定义方法。

```c
#include<stdio.h>
#include<string.h>
struct _INFO
{
    int num;
    char str[256];
};
int main()
{
    struct _INFO A;
    A.num = 31;
    printf("This day is %d \n", A.num, A.str);
    return 0;
}
```

输出结果如图 2-17 所示。

图 2-17 结构体的定义

这个程序定义了结构体_INFO,将 A 说明为_INFO 类型的结构变量,对其赋值 31,并输出结果。结构体除了由多个数据成员构成,还可以由多个数组成员构成。结构体数组的每一个元素都是结构体类型。

下面我们通过一个例子来学习一下结构体数组的使用。

```c
#include<stdio.h>
#define NUM 3//定义后续使用的 NUM 值为 3
int main()
{
 struct students
 {
  char name[20];
  int age;
 };
struct students stu[NUM] = {{"A",18},{"B",19},{"C",18}};
printf("姓名    年龄\n\n");
for(int i = 0;i<NUM;i++)
{
printf("%s    %d\n\n",stu[i].name,stu[i].age);
}
return 0;
}
```

输出结果如图 2-18 所示。

图 2-18 结构体数组

这个程序实现了表示学生的名字和年龄的功能。每一个学生的名字和年龄的数据都是一个数组,该结构体数组由这三个数组成员组成。

当我们需要将几种不同类型的变量存放同一段单元中时,我们就要选择使用共用体。共用体的作用是允许几种不同类型的变量存放到同一段内存单元中,几种变量互相覆盖。

我们在使用共用体时应注意共用体在每个瞬间只能有一个成员起作用,其他成员不起作用,且共用体中起作用的成员是最后一次存放的成员,在存入一个新成员后,原有成员就失去作用。

下面我们通过一个例子来学习一下共用体的使用。

```
#include<stdio.h>
union INFO
{
    int a;
    int b;
    int c;
};
int main()
{
    union INFO A;
    A.a = 1;
    A.b = 2;
    A.c = 3;
    printf("a:%d\n",A.a);
    printf("b:%d\n",A.b);
    printf("c:%d\n",A.c);
    return 0;
}
```

输出结果如图 2-19 所示。

图 2-19　共用体的定义和使用

这是由于共用体变量在一瞬间只能够有一个成员起作用,且起作用的为最后存放的成员 3,故前面存放的 1 和 2 失去作用,故最终输出结果都为 3。

当需要使用新的数据类型名称或结构体时,我们可以使用类型说明语句 typedef 定义新的类型来代替已有的类型。typedef 最常用的作用就是给结构体变量重命名。

typedef 语句的一般形式是:

typedef 已定义的类型 新的类型;

下面我们通过一个例子来学习一下 typedef 语句的使用。

```
#include<stdio.h>
#include<string.h>
typedef struct _INFO
{
    int num;
    char str[256];
}INFO;
int main()
{
    struct _INFO A;
    INFO B; //通过 typedef 重命名后的名字 INFO 与 struct _INFO 完全等价!
    A.num = 18;
    B = A;
    printf("Your age is %d\n", A.num);
    printf("Your age is %d\n", B.num);
    return 0;
}
```

输出结果如图 2-20 所示。

图 2-20　typedef 的使用

从上面的例子可以 typedef 能为关键词改名,使改名之后的 INFO 类型等价于 struct_INFO 类型,让我们在定义这种结构类型时更方便、省事。

思 考 题

1. 输出 My name is 2023!

这个例子让计算机打印出相关信息。

```
printf("My name is 2023! \n");
```

首先,printf 是一个 C 语言的函数,打印出信息。后面必须用()来完成内容。

My name is 2023! 是我们要输出的内容。

\n 是计算机显示时自动回车换行。

这里用到三个常用的 DOS 命令:

```
cd 切换路径
dir 查看当前文件
cd.. 退回到上一级目录
```

2. 设计一个带加、减、乘、除功能的计算器。

在 C 语言中,有参数和值的概念,我们来设计一个简单的计算器。

```
#include<stdio.h>
int main()
{
    int x = 1;
    int y = 2;
    int z = x + y;
    printf("z = %d \n",z);
    return 0;
}
```

学会一个重复的命令操作:用鼠标选择命令行界面后,键盘的上下键就可以重复以前输入的命令。

注意,按下键盘中的窗口键+R 键后,会出现一个小窗口,输入"cmd"后按回车键,就可以输入命令。这个"dir"是以前输入的,这个"dir",是采用键盘的上下箭头操作,重复以前的命令得到的。

3. 完成 1+2+3+4+……100 的和。

编写程序代码如下:

```
#include<stdio.h>
int main()
{
int sum = 0;
    for(int i = 1; i<101; i++)
    {
```

```
        sum += i;
    }
        printf("sum = %d \n",sum);
    return 0;
}
```

程序说明:

```
int sum = 0;     是定义一个整数型变量 sum,并且给这个变量赋值为 0
for(int i = 1; i < 101; i++)
这个是一个循环结构 for,
用于控制循环的起点,int i = 1
用于控制循环的终点,i < 101
用于控制步长,i++ 每次循环,添加 1
{
    sum += i;
    赋值语言,是从右往左赋值。
    这句话的意思是,将 i 的值取出后,于 sum 当前的值相加后的结果,再重新赋给了 sum。
}
printf("sum = %d \n",sum);
%d 是输出为整数格式
\n 是回车换行。
return 0;
```

4. 输入你的名字,可以打印出你的名字,你父亲的名字,你母亲的名字。

程序代码编写如下:

```
#include "stdafx.h"

struct family{
    char * myname;
    int    myage;
    char * myfather;
    char * mymother;
};
int _tmain(int argc, _TCHAR * argv[])
{
    struct family myfamily;
    myfamily.myname = "walter";
    myfamily.myage = 25;
    myfamily.myfather = "father";
    myfamily.mymother = "mother";
    if(myfamily.myname == "walter")
    {
        printf("myname is %s, I am %d old, myfater = %s, mymother = %s \n",myfamily.myname,
            myfamily.myage,myfamily.myfather,myfamily.mymother);
    }
```

```
    return 0;
}
```

程序说明

```
#include "stdafx.h"
```
引用你所需要的头文件。
```
struct family{
    char * myname;
    int    myage;
    char * myfather;
    char * mymother;
};
```

这个是一个结构体的定义,用关键字 struct 作为结构体的定义。

```
int _tmain(int argc, _TCHAR * argv[])
{
    struct family myfamily;
```
定义了一个变量 myfamily,它们都是 family 类型,都由 5 个成员组成。
```
    myfamily.myname = "walter";
    myfamily.myage = 25;
    myfamily.myfather = "father";
    myfamily.mymother = "mother";
```
给这个变量进行赋值。
```
    if(myfamily.myname == "walter")
```
比较结构体的值和当前的值,是否相等。
```
    {
        printf("myname is %s, I am %d old, myfater=%s, mymother=%s \n",myfamily.myname,
        myfamily.myage,myfamily.myfather,myfamily.mymother);
    }
```
打印出结果,

%s 输出字符串,

%d 输出整数。

```
    return 0;
}
```

第 3 章

C++语言

C++语言是美国贝尔实验室 Bjarne Stroustrup 博士在 C 语言的基础上开发而成的,其主要目的是解决 C 语言的三大问题,即 C 语言类型检测相对较弱、C 语言缺乏支持代码重用的语言结构和 C 语言不适合开发大型程序。总体来看,C 语言和 C++语言主要有四个不同之处:第一,C 语言用的 malloc()和 free()分别用于动态内存的申请和释放,C++语言提供 new 和 delete,分别用于内存的申请和释放。第二,同时 C++语言引入内联函数 inline,这样做的目的是当调用内联函数时候,C++语言编译器将使用函数体种的代码替代函数调用表达式(加快代码的执行,减少调用开销)。第三,C++语言有函数重载的功能,然而 C 语言要求函数名必须是唯一的。第四,C++语言的应用,其功能类似为变量起一个别名,主要应用于函数参数及函数的返回值。

3.1 类和对象

C 语言是将数据与处理方法分开,然而 C++语言面向对象的编程是将数据与处理方法封装为类。在 C++语言中,对象的类型就称为类,即类代表了某一批对象的共性和特征。

```
Class <类名>
{
private:只是本类调用
public:共有数据成员类内和类外的函数访问或调用。
protected:本类和本类派生的成员函数访问或调用。
}
```

其中,"∷"是 C++语言中新引入的运算符,称为"作用域运算符"。

举一个例子:

```
#include <iosteam.h>
float x = 10;
void main()
{
    int x = 100;
```

```
cout << x << std :: endl;
}
```

那么这个输出是什么?

下一个例子:

```
#include<iosteam.h>
float x = 10;
void main()
{
    int x = 100;
    cout << :: x << std :: endl;
}
```

这个输出的结果又是如何呢?

类是一种数据类型,因此用类也可以定义变量。用类定义的变量称为对象。总之,对象是类的实例,是属于某个已知的类,同时,对象必须先定义,再使用。

下面来介绍构造函数与析构函数。首先,这两个函数不能有返回类型,也不能是 void 类型。一个类中可以有多个参数或者参数类型不同的构造函数。析构函数以调用构造函数相反的顺序被调用。

静态成员:声明为 static 的类成员便能在类范围中共享,称之为静态成员。全局变量给面向对象程序带来的问题就是违背封装原则。如果要使用静态数据成员,必须在程序运行之前分配空间和初始化。

类的封装固然好,但是有时需要有一种机制来访问类的所有函数,所以就产生了友元(用 friend 关键字)。友元是一种定义在类外部的普通函数,但是它需要在类体内进行说明。友元函数不是成员函数,它是类的朋友,所以可以访问类的全部成员。

3.2 类的继承

在 C++语言中,基类的共有成员默认情况下在派生类中是私有的,而私有成员在派生类的成员函数是不能被访问的。派生新类需要三个步骤:吸收基类成员、改造基类成员和添加新的成员。下面详细介绍这三个步骤。

1. 吸收基类成员

在 C++语言的类继承中,首先是将基类的成员全盘接收,这样派生类实际上就包含了它的所有基类的除构造函数和析构函数之外的所有成员。

2. 改造基类成员

对于基类成员的改造包括两个方面:一个是基类成员的访问控制问题,主要依靠派生类定义的继承方式来控制;二是对基类数据或者函数成员的覆盖,就是在派生类中定义一个和基类数据或者函数同名的成员,由于作用域不同,于是发生同名覆盖,基类的成员就被替换成派生类中的同名成员。

3. 添加新的成员

派生类新成员的加入是继承与派生机制的核心,是保证派生类在功能上有所发展的关键。我们可以根据实际情况,给派生类添加适当的数据和函数成员以实现必要的新增功能。同时,在派生过程中,基类的构造函数和析构函数是不能被继承下来的。在派生类中,一些特别的初始化和扫尾清理工作也需要我们重新加入新的构造函数和析构函数。

3.3 类的多态

所谓多态,就是不同对象收到相同的信息时产生的动作。也就是说,多态是指用一个名字定义不同的函数,这些函数执行不同但又类似的操作,即用同样的访问功能不同的函数,从而实现"一个接口,多种方法"。

C++一个源程序讲过编译,连接,成为可执行文件的过程,是把可执行代码联编在一起的过程。其中,把在运行之前就完成的联编成为静态联编;二在程序运行是才完成的联编叫作动态联编。

静态联编是指系统在编译时就决定如何实现某一动作。静态联编要求在程序编译时就知道调用函数的全部信息。因此,这种联编类型的函数调用速度很快,效率高是静态联编的主要优点。

动态联编是指系统在运行是动态实现某一动作。采用这种联编方式,一直要到程序运行是才能确定调用哪个函数。动态联编的主要优点是:提供了更好的灵活性,问题抽象性和程序易维护性。

纯虚函数的一般形式如下:

```
virtual type func_name(parameter) = 0;
```

如果一个类至少有一个纯虚函数,那么就称该类为抽象类。因此,对于抽象类的使用有如下规定:

1. 由于一个抽象类中至少包含有一个没有定义功能的纯虚函数,因此抽象类只能用作其他类的基类,不能建立抽象类对象。

2. 抽象类不能用作参数类型,函数返回类型或者显式转换的类型,但可以声明指向抽象类的指针或引用,此指针可以指向它的派生类,进而实现多态。

3.4 类的模板

模板也叫参数化的类型,是实现类属机制的一种工具。模板的功能非常强大,可以有效提高程序设计的效率。

比如,我们可以定义类的模板如下:

```
template< class T>
```

其中,class 和类定义的 class 不同,表示参数 T 是一个数据类型。在使用该模板时,T 可以用用户定义的数据类型,也可以用 C++语言固有的数据类型,如 int,float 等替换。

举一个例子:用函数模板求较大值。

```
template<class T>
T max(T x,T y)
{
return(x>y)? x:y;
}
```

最后举一个完整的例子:用冒泡法求最大值。

```
#include"iostream.h"
template<class T>//T 是向量中元素的数据类型
void Bubblesort(T a[],int n)//Bubblesort 函数模板
{
int i,j;
T temp;
for(j=0;j<n;j++)
for(i=0;i<n-1-j;i++)
if(a[i]>a[i+1])
{
temp=a[i];
a[i]=a[i+1];
a[i+1]=temp;
}
}
template<class T>
void disp(T a[],int n)//disp 函数模板
{
for(int i=0;i<n;i++)
cout<<a[i]<<"";
cout<<endl;
}
void main()
{
int a[]={3,8,2,6,7,1,4,9,5,0};
char b[]={'i','d','a','7',.b','f','e','C','9','h'};
cout<<"整数排序:"<<endl;
cout<<"原序列:"11 ;
disp(a,10);
Bubblesort(a,10);
cout<<"新序列:勺
disp(a,10);
```

```
    cout << "字符排列:" << endl;
    cout << "原序列:,';
    disp(b,10);
    Bubblesort(b,10);
    cout << "新序列:11 ;
    disp(b,10);
}
```

这个冒泡的程序,采用类模板的方法完成,是掌握类模板用法的一个比较典型的例子。

C++于C语言最大的区别是采用了类的设计,就像我们人类繁衍一样,我们的DNA中继承了父母的相当一部分基因,然后,我们又有自己的一些特点。同时,我们的兄弟姐妹之间,又有一些关系。

C++一个重要的概念是类,其最重要的特点是继承与多态。就像我们每个人继承了父母的一些基因一样,同时,我们兄弟姐妹又各不相同的,在C++定义为,多态。

两个最重要的名词:虚函数和纯虚函数。

虚函数,是可以根据后期绑定,来执行最后的结果。

纯虚函数,如果只是定义一个没有实际实现的函数,可以在类中定义一个纯虚函数,以便在子类中,各自实现有实际意义的操作和功能。

这里,我给大家重点讲解一下标准模板库中的vector。

作为一个C++程序员,很多时候,都需要用到数据的存储和过滤,这个时候,我们可以考虑选择用vector作为我们的存储数据的方法。

比如,我们需要设计一个数据结构,将我们全班同学的姓名,学号,语文,数学,外语三门成绩进行统计,并按照总分从高到低进行排名。

具体的代码如下:

```cpp
#include "pch.h"
#include <iostream>

#include <vector>
struct student {
    char * name;
    char * id;
    float chinese;
    float math;
    float english;
    float total;
};
int number = 3;


int main()
{
    std :: vector < student > students;
```

```cpp
    struct student student1, student2, student3;
    student1.name = "aaa";
    student1.id = "2020001";
    student1.chinese = 75.7;
    student1.math = 80;
    student1.english = 90;
    student1.total = student1.chinese + student1.math + student1.english;
    students.push_back(student1);

    student2.name = "bbb";
    student2.id = "2020002";
    student2.chinese = 93.5;
    student2.math = 99;
    student2.english = 98;
    student2.total = student2.chinese + student2.math + student2.english;
    students.push_back(student2);

    student3.name = "ccc";
    student3.id = "2020003";
    student3.chinese = 55.5;
    student3.math = 61;
    student3.english = 72;
    student3.total = student3.chinese + student3.math + student3.english;
    students.push_back(student3);
    int maxtotal = 0;
    float nummax = 0;
    for(int i = 0; i < students.size(); i++)
    {
        if(students[i].total > maxtotal)
        {
            maxtotal = students[i].total;
            nummax = i;
        }
    }
    std::cout << "The best student name is " << students[nummax].name << std::endl;
    std::cout << "The best student total is " << students[nummax].total << std::endl;
    std::cout << "The best student is " << students[nummax].name << " chinese is " << students[nummax].chinese << " math is " << students[nummax].math << " english is " << students[nummax].english << std::endl;
}
```

运行结果如图 3-1 和图 3-2 所示。

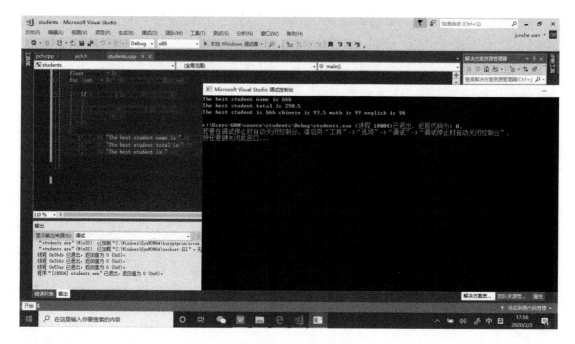

图 3-1　源代码运行的结果

图 3-2　注意设定的参数

思　考　题

一、简答题

1. 类和对象之间的区别是什么？
2. 编译时的多态性与运行时的多态性有什么区别？它们的实现方法有什么不同？

二、选择题

1. 关于函数模板,以下哪句话描述是正确的?(　　)

A. 函数模板是一个特殊的函数。

B. 函数模板定义中,只能包含一个参数。

C. 通过函数模板可以生成具体的函数。

D. 函数模板中每个参数在函数参数表中不一定非要使用一次。

2. 在程序中使用模板,以下哪句话是错误的?(　　)

A. 模板函数是由函数模板实例化而来的。

B. 通过给出实参,可以从类模板生成具体的类。

C. 使用模板可以简化程序的设计,减少重复编码。

D. 不能在程序中定义模板类的对象。

三、应用题

1. 编写程序。定义求绝对值的函数模板 abs,并在 main 函数中由此模板生成三个模板函数,分别计算整数、浮点数、双精度浮点数的绝对值。

2. 定义一个栈的类模板 stack 包括两个私有数据成员 data[ ](存放栈中的元素)和 top(栈顶元素的下标),以及两个公有的成员函数 push(入栈)和 pop(出栈),并由此模板生成两个模板类:整数栈和字符栈。

第 4 章
CMake 构建系统

　　CMake 是一个跨平台的安装(编译)工具,可以用简单的语句来描述所有平台的安装(编译过程),CMake 的组态档取名为 CMakeLists.txt,其主要目的是方便系统的配置相关头文件和库的时候,可以做到快速配置。本系统采用 Win 10(64 位),CMake 3.22.1,VisualStuio 2019,QT 5.15.0,openCV 4.5.3,cuda 10.1,Tensorflow 2.3.0,编辑采用 Visual Studio Code 1.61.0。本实验需要安装以上软件后,才能进行下面的实验。

4.1　用 CMake 构建一个 HelloWorld

　　第一,采用 Visual Studio Code 编辑一个 hello world.cpp 文件如下。

```
#include<stdio.h>
int main()
{
    fprintf(stdout,"Hello world!\n");
    return 0;
}
```

　　第二,同样采用 Visual Studio Code 编辑一个 CMakeLists.txt 文件后,如图 4-1 所示。

```
cmake_minimum_required(VERSION 2.8)
project(Helloworld)
add_executable(Helloworld hello world.cpp)
```

　　第三,选中 CMakeLists.txt 后,选择打开方式如图 4-1 所示。
　　选择 cmake.exe,如图 4-2 所示。
　　第四,点击确定后,会生成如下文件,如图 4-3 所示。
　　第五,双击 Helloworld.sln,用 VisualStudio2019 打开后,如图 4-4 所示。再选择"重新生成",生成如图 4-5 所示。

第 4 章 CMake 构建系统

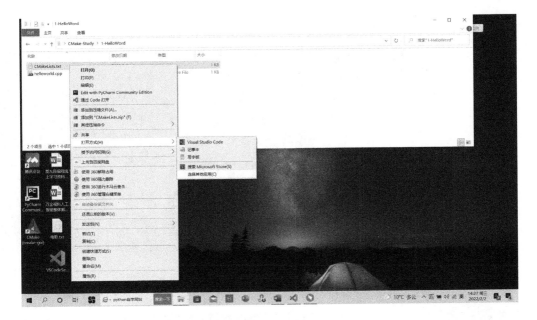

图 4-1 选择指定的 CMakeList.txt 文件

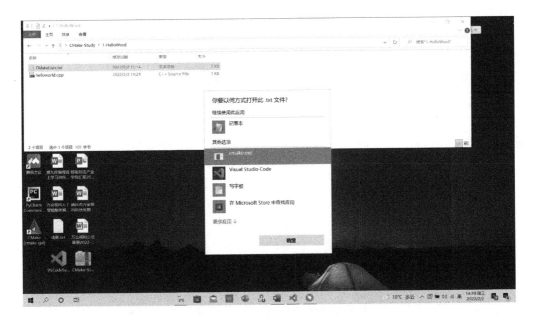

图 4-2 使用 cmake.exe 进行打开

图 4-3 会生成一个 Helloworld.sln 文件

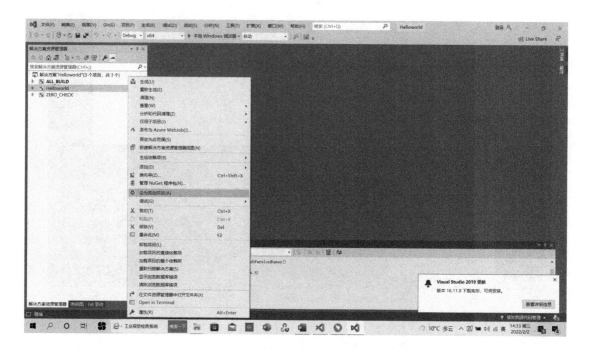

图 4-4　使用 VS2019 打开 Helloworld.sln 工程文件

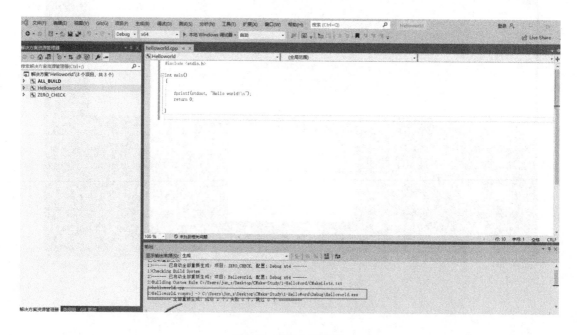

图 4-5　重新生成代码

第六，点击启动，结果如图 4-6 所示。本次实验使用 CMake 的方法，采用 Visual Studio Code 进行编辑，采用 cmake.exe 进行编译，最后采用 Visual Studio 2019 进行运行及得到结果。

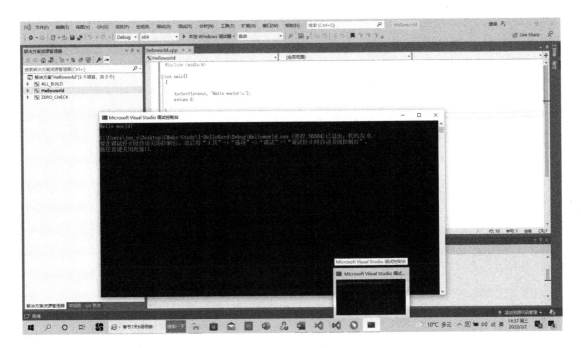

图 4-6 运行后的结果

4.2 CMake 与 OpenCV

有了上面的基础,我们就开始用 CMake 对 OpenCV 的环境进行配置。首先需要对配置文件"CMakeList.txt"进行设计,其详细内容如下。

```
cmake_minimum_required(VERSION3.12)

# set PROJECT_NAME and version
project(HelloWorld)
set(VERSION_MAJOR0)
set(VERSION_MINOR0)
set(VERSION_PATCH1)

# set path for OpenCV
set(OpenCV_DIR C:/opencv/build)

# without console by default
# set(CMAKE _ EXE _ LINKER _ FLAGS" $ { CMAKE _ EXE _ LINKER _ FLAGS } /SUBSYSTEM:WINDOWS /ENTRY:
mainCRTStartup")

# Find includes in corresponding build directories
set(CMAKE_INCLUDE_CURRENT_DIR ON)
```

```cmake
# Instruct CMake to run moc automatically when needed
set(CMAKE_AUTOMOC ON)
# Create code from a list of Qt designer ui files
set(CMAKE_AUTOUIC ON)

# openCV add by wjs
find_package(OpenCV REQUIRED)

file(GLOB SOURCE_FILE *.cpp)
file(GLOB HEAD_FILE *.h)

set(project_headers
    ${HEAD_FILE})

set(project_sources
    ${SOURCE_FILE})

add_executable(${PROJECT_NAME} ${project_headers} ${project_sources})

# Use openCV
target_link_libraries(${PROJECT_NAME} ${OpenCV_LIBS})
```

Helloworld.cpp 内容如下。

```cpp
#include<opencv2/opencv.hpp>
#include<iostream>
#include<string>

using namespace std;
using namespace cv;

int main(int argc, char **argv)
{
    string file = "./pics/Lena.jpg";

    image = imread(file, 1);
    if(!image.data)
    {
        cout<<"No image data"<<endl;
    return -1;
    }
    imshow("color", image);
    cv::waitKey(0);

    Mat gray_img;
```

```
    cvtColor(image, gray_img, COLOR_BGR2GRAY);
    imshow("gray", gray_img);
    cv::waitKey(0);

    return 0;
}
```

采用 cmake.exe 进行构建后，结果如图 4-7 所示。

图 4-7　运行时缺少 opencv_world453d.dll

将 OpenCV 下面对应的 dll 拷贝到 debug 目录下后，彩色图像和灰度化后的图像对比效果如图 4-8 所示。

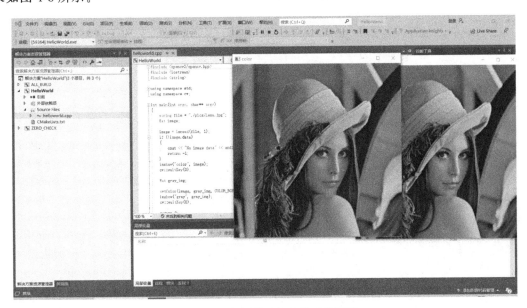

图 4-8　彩色图像和灰度化后的图像

4.3　CMake 与 QT

首先,我们来看看下面的 CMakeLists.txt。

```
cmake_minimum_required(VERSION3.12.0)

# set PROJECT_NAME and version
project(HelloWorld)
set(VERSION_MAJOR 0)
set(VERSION_MINOR 0)
set(VERSION_PATCH 1)

# set path for QT
set(CMAKE_PREFIX_PATH C:/QT/5.15.0/msvc2019_64)

# without console by default
set(CMAKE_EXE_LINKER_FLAGS" ${CMAKE_EXE_LINKER_FLAGS} /SUBSYSTEM:WINDOWS /ENTRY:mainCRTStartup")

# Find includes in corresponding build directories
set(CMAKE_INCLUDE_CURRENT_DIR ON)
# Instruct CMake to run moc automatically when needed
set(CMAKE_AUTOMOC ON)
# Create code from a list of Qt designer ui files
set(CMAKE_AUTOUIC ON)

find_package(Qt5 REQUIRED COMPONENTS Core Widgets Gui)

set(log4cplus_DIR ${PROJECT_SOURCE_DIR}/lib/log4cplus/lib/cmake/log4cplus)
find_package(log4cplus REQUIRED)
if(log4cplus_FOUND)
    message("found log4cplus ")
endif()

file(GLOB SOURCE_FILE *.cpp)
file(GLOB UI_FILE *.ui)
file(GLOB HEAD_FILE *.h)

set(project_ui
    ${UI_FILE})

set(project_headers
```

```
    ${HEAD_FILE})

set(project_sources
    ${SOURCE_FILE})

add_executable( ${PROJECT_NAME} ${project_headers} ${project_ui} ${project_sources})

# Use the widgets module from Qt5
target_link_libraries( ${PROJECT_NAME}
    Qt5::Core
    Qt5::Gui
    Qt5::Widgets)

target_link_libraries( ${PROJECT_NAME} log4cplus::log4cplusU)

set(CMAKE_INSTALL_PREFIX ${PROJECT_BINARY_DIR}/installed)
set(WINDEPLOYQT_DIR C:/QT/5.15.0/msvc2019_64/bin)
# exec windeployqt automatically when build install in VS
install(CODE "execute_process(COMMAND ${WINDEPLOYQT_DIR}/windeployqt.exe
              ${PROJECT_BINARY_DIR}/Release/${PROJECT_NAME}.exe)")
install(DIRECTORY ${PROJECT_BINARY_DIR}/Release/ DESTINATION /)

install(FILES ${PROJECT_SOURCE_DIR}/Mylog.conf DESTINATION /)
install(FILES ${PROJECT_SOURCE_DIR}/lib/log4cplus/bin/log4cplusU.dll DESTINATION .)
```

然后,构建一个 build.bat。

```
@echo off

:: delete build dir
if "%1" == "delete" goto _DEL

if not exist build md build

cd build
cmake -A x64 -DCMAKE_CONFIGURATION_TYPES=Release ..
cd ..
goto _END

:_DEL
rd build /s /q
```

```
goto _END

:_END
```

接着,看看程序的头文件和实现文件的整体结构,如图 4-9 所示。

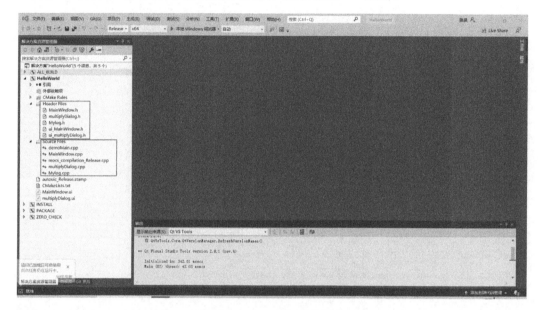

图 4-9　VS2019 构建的源码系统

最后,结果如图 4-10 所示。

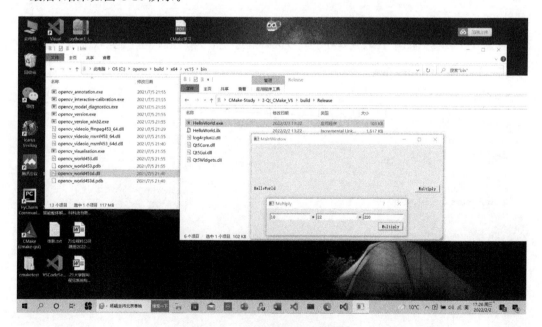

图 4-10　一个乘法器运行的结果

本次实验是通过建立 CMakeLists.txt 文件,建立批处理文件,生成 Visual Studio Code 的工程文件,实现一个加法的过程。

4.4　CMake 与 Tensorflow

首先,构建一个 CMakeLists.txt。

```
cmake_minimum_required(VERSION3.12.0)

# set PROJECT_NAME and version
project(HelloWorld)
set(VERSION_MAJOR0)
set(VERSION_MINOR0)
set(VERSION_PATCH1)

# set path for QT
set(CMAKE_PREFIX_PATH C:/QT/5.15.0/msvc2019_64)

# set path for OpenCV
set(OpenCV_DIR C:/opencv/build)

# set path for Tensorflow
set(TF_INCLUDE C:/libtensorflow-gpu-windows-x86_64-2.3.0/include)
set(TF_LIB C:/libtensorflow-gpu-windows-x86_64-2.3.0/lib)

# set path for balser camera
set(BALSER_INCLUDE C:/Users/jun_s/Desktop/cmaketest/Qt_CMake_VS-OpenCV-TF/balser/include)
set(BALSER_LIB C:/Users/jun_s/Desktop/cmaketest/Qt_CMake_VS-OpenCV-TF/balser/lib/x64)

# without console by default
set(CMAKE_EXE_LINKER_FLAGS" ${CMAKE_EXE_LINKER_FLAGS} /SUBSYSTEM:WINDOWS /ENTRY:mainCRTStartup")

# Find includes in corresponding build directories
set(CMAKE_INCLUDE_CURRENT_DIR ON)
# Instruct CMake to run moc automatically when needed
set(CMAKE_AUTOMOC ON)
# Create code from a list of Qt designer ui files
set(CMAKE_AUTOUIC ON)

find_package(Qt5 REQUIRED COMPONENTS Core Widgets Gui SerialPort Multimedia)
# openCV add by wjs
find_package(OpenCV REQUIRED)
```

```cmake
# Tensorflow add by wjs
include_directories(${TF_INCLUDE})
link_directories(${TF_LIB})
# BALSER add by wjs
include_directories(${BALSER_INCLUDE})
link_directories(${BALSER_LIB})

set(log4cplus_DIR ${PROJECT_SOURCE_DIR}/lib/log4cplus/lib/cmake/log4cplus)
find_package(log4cplus REQUIRED)
if(log4cplus_FOUND)
    message("found log4cplus ")
endif()

file(GLOB SOURCE_FILE *.cpp)
file(GLOB UI_FILE *.ui)
file(GLOB HEAD_FILE *.h)

set(project_ui
    ${UI_FILE})

set(project_headers
    ${HEAD_FILE})

set(project_sources
    ${SOURCE_FILE})

add_executable(${PROJECT_NAME} ${project_headers} ${project_ui} ${project_sources})

# Use the widgets module from Qt5
target_link_libraries(${PROJECT_NAME}
    Qt5::Core
    Qt5::Gui
    Qt5::Widgets
    Qt5::SerialPort
    Qt5::Multimedia)
# Use log4cplus and openCV
target_link_libraries(${PROJECT_NAME} log4cplus::log4cplusU ${OpenCV_LIBS})

# Use Tensorflow
target_link_libraries(${PROJECT_NAME} tensorflow.lib)

# Use BALSER
```

```cmake
target_link_libraries( ${PROJECT_NAME} GCBase_MD_VC141_v3_1_Basler_pylon.lib GenApi_MD_VC141_v3_1_Basler_pylon.lib PylonBase_v6_2.lib PylonC.lib PylonGUI_v6_2.lib PylonUtility_v6_2.lib)

set(CMAKE_INSTALL_PREFIX ${PROJECT_BINARY_DIR}/installed)
set(WINDEPLOYQT_DIR C:/QT/5.15.0/msvc2019_64/bin)
# exec windeployqt automatically when build install in VS
install(CODE"execute_process(COMMAND ${WINDEPLOYQT_DIR}/windeployqt.exe
             ${PROJECT_BINARY_DIR}/Release/${PROJECT_NAME}.exe)")
install(DIRECTORY ${PROJECT_BINARY_DIR}/Release/ DESTINATION /)

install(FILES ${PROJECT_SOURCE_DIR}/Mylog.conf DESTINATION /)
install(FILES ${PROJECT_SOURCE_DIR}/lib/log4cplus/bin/log4cplusU.dll DESTINATION .)

set(CPACK_GENERATOR NSIS)
set(CPACK_PACKAGE_NAME ${PROJECT_NAME})
set(CPACK_PACKAGE_VERSION_MAJOR ${VERSION_MAJOR})
set(CPACK_PACKAGE_VERSION_MINOR ${VERSION_MINOR})
set(CPACK_PACKAGE_VERSION_PATCH ${VERSION_PATCH})
set(CPACK_PACKAGE_INSTALL_DIRECTORY ${PROJECT_NAME})

INCLUDE(CPack)
```

然后,构建一个 build.bat。

```bat
@echo off

:: delete build dir
if "%1" == "delete" goto _DEL

if not exist build md build

cd build
cmake -A x64 -DCMAKE_CONFIGURATION_TYPES=Release ..
cd ..
goto _END

:_DEL
rd build /s/q
goto _END

:_END
```

构建整个系统的过程如图 4-11 所示。

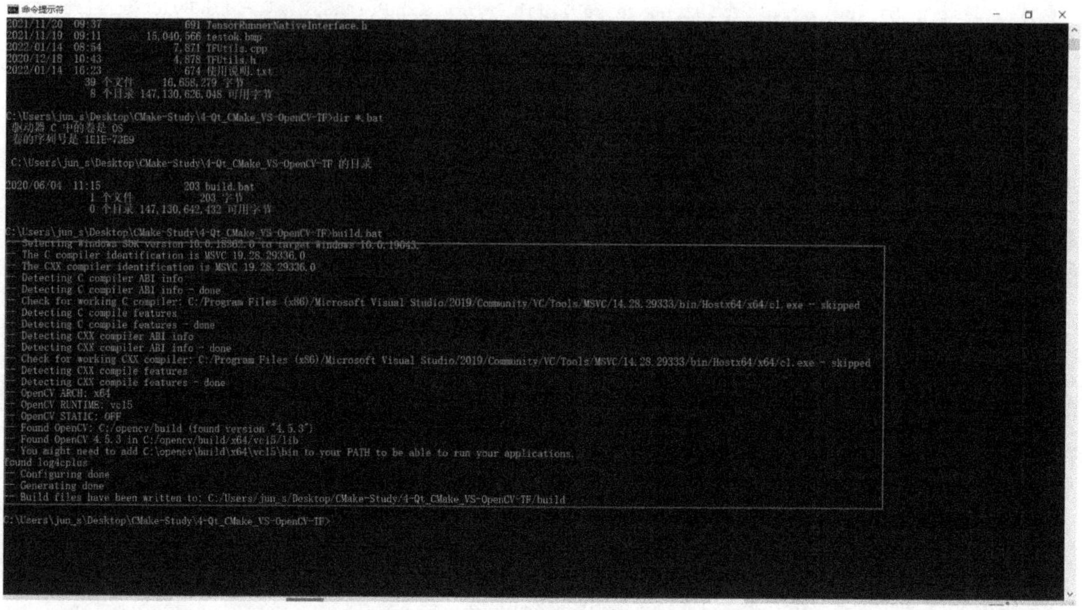

图 4-11 构建整个系统的过程

在 Qt_CMake_VS-OpenCV-TF 目录下，点击 build.bat 会自动生成 HelloWorld.sln 如图 4-12 所示。

图 4-12 使用 build.bat 构建工程文件 HelloWorld.sln

选择工程文件 HelloWorld.sln，如图 4-13 所示。

将版本默认的 C++14 改为 C++17，如图 4-14 所示。

在编译器默认的情况下采用 C++14，由于程序会用到 filesystem 连续读取图片，所以必须将默认的 C++14 改为 C++17 版本后，重新编译。结果如图 4-15 所示。

第 4 章 | CMake 构建系统

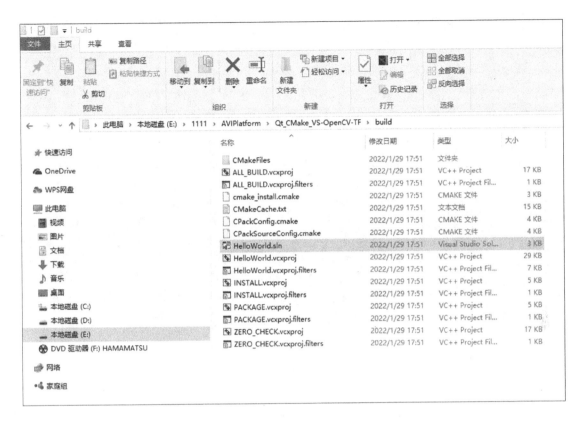

图 4-13　选择工程文件 HelloWorld.sln

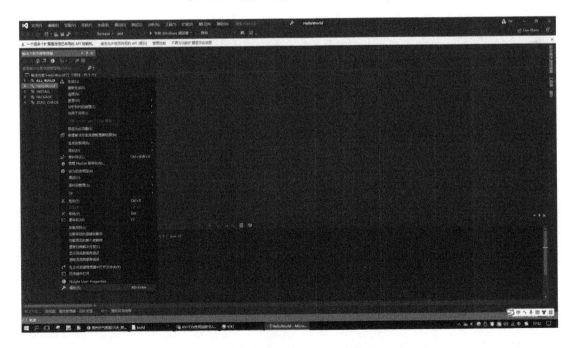

图 4-14　将版本默认的 C++14 改为 C++17

C++语言与机器视觉编程实战

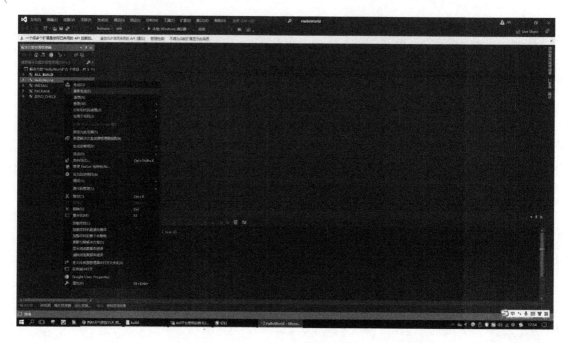

图 4-15 选择重新编译

将 must 目录的文件拷贝到 Release 下，如图 4-16、图 4-17 和图 4-18 所示。

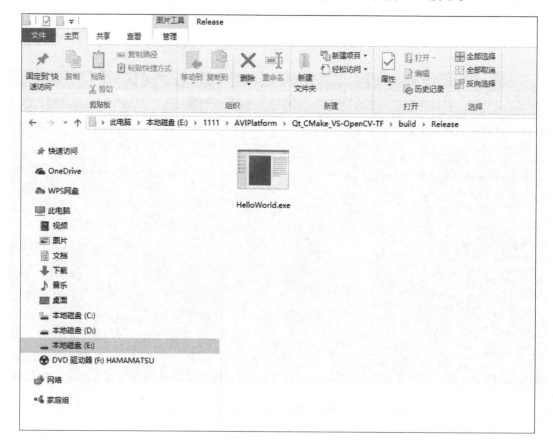

图 4-16 编译生成可执行文件 HelloWorld.txt

第 4 章 CMake 构建系统

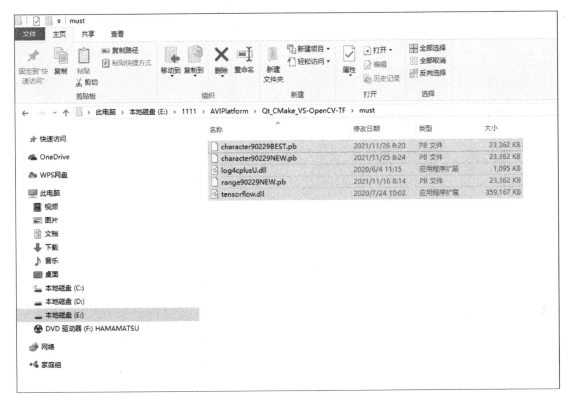

图 4-17 拷贝相关的文件及动态库

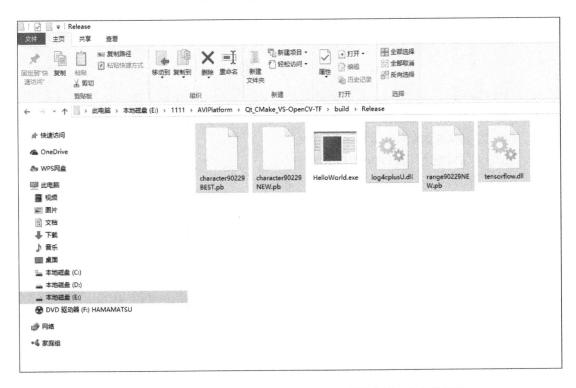

图 4-18 将相关文件拷贝到 HelloWorld.exe 所在的同一个文件目录

运行 HelloWorld.exe,结果如图 4-19 所示。

图 4-19 运行后的结果图

本系统采用 Win 10(64 位),CMake 3.22.1,Visual Studio 2019,QT 5.15.0,openCV 4.5.3,cuda 10.1,Tensorflow 2.3.0,编辑采用 Visual Studio Code 1.61.0。本实验需要安装以上软件后,才能进行下面的实验。

思 考 题

简答题

1. CMake 如何配置 Release 版本?
2. CMake 如何配置 OpenCV 图像开发库?
3. CMake 如何配置 Basler 相机的头文件和库文件?

第 5 章 QThread 多线程

在 C++编程技巧中,多线程编程是必须掌握的内容。在 QT 编程的框架下采用 QThread 构建的多线程方式是一个比较好的选择。本章通过一个骰子游戏中的多线程实现来学习 QThread 多线程技术。本章 5.1 节介绍了使用 QThread 类来实现多线程,5.2 节介绍了采用 QMutex 和 QMutexLocker 实现多线程,5.3 节介绍了采用定时器触发方式实现多线程,5.4 节介绍了采用生成者和消费者模式实现多线程。

5.1 基于 QThread 多线程

首先,利用 QThread 多线程设计能达到如图 5-1 所示的结果的程序。

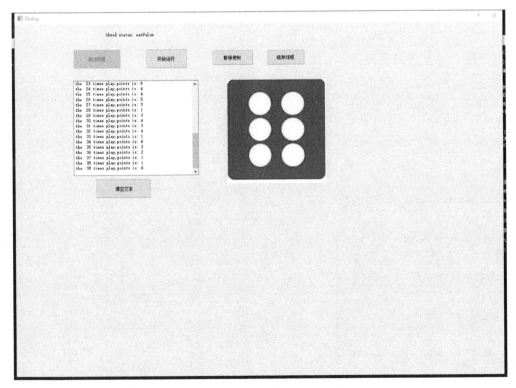

图 5-1 多线程设计的目标显示

首先是线程的 H 文件设计。

```cpp
#ifndef QDICETHREAD_H
#define QDICETHREAD_H

#include <QObject>
#include <Qthread>
class qdicethread: public QThread
{
    Q_OBJECT
private:
    int m_seq = 0;
    int m_diceValue;
    bool m_Paused = true;
    bool m_stop = false;
protected:
    void run() Q_DECL_OVERRIDE;
public:
    qdicethread();
    void diceBegin();
    void dicePause();
    void stopThread();
signals:
    void newValue(int seq, int diceValue);
};
#endif // QDICETHREAD_H
```

然后是 CPP 文件的设计。

```cpp
#include "qdicethread.h"
#include <Qtime>

qdicethread :: qdicethread()
{
}
void qdicethread :: diceBegin()
{
    m_Paused = false;
}
void qdicethread :: dicePause()
{
    m_Paused = true;
}
void qdicethread :: stopThread()
{
```

```cpp
    m_stop = true;
}
void qdicethread :: run()
{
    m_stop = false;
    m_seq = 0;
    QTime * time = newQTime();
    QString strTime;
    qsrand(time -> currentTime().msec());
    while(!m_stop)
    {
        if(!m_Paused)
        {
            m_diceValue = qrand();
            m_diceValue = (m_diceValue % 6) + 1;
            m_seq ++ ;
            emitnewValue(m_seq, m_diceValue);
        }
        msleep(500);
    }
    quit();
}
```

接着是对话框的 H 文件的设计。

```cpp
#ifndef DICEDIALOG_H
#define DICEDIALOG_H

#include <QDialog>
#include "qdicethread.h"
namespace Ui {
class diceDialog;
}
class diceDialog: public QDialog
{
    Q_OBJECT
private:
    qdicethread threadA;
protected:
    void closeEvent(QCloseEvent * event);
public:
    explicitdiceDialog(QWidget * parent = nullptr);
    ~diceDialog();
private slots:
```

```
        void onthreadA_started();
        void onthreadA_finished();
        void onthreadA_newValue(int seq, int diceValue);
        void on_btnEnable();
        void on_btnStart();
        void on_btnPause();
        void on_btnStop();
        void on_btnClear();
private:
        Ui::diceDialog * ui;
};
#endif // DICEDIALOG_H
```

最后是对话框的 CPP 设计。

```
#include "diceDialog.h"
#include "ui_diceDialog.h"
#include <QImage>
#include <QDir>
#include <QLibrary>
#include <QString>
#include <QMessageBox>


diceDialog::diceDialog(QWidget * parent):
    QDialog(parent),
    ui(new Ui::diceDialog)
{
    ui->setupUi(this);
    QObject::connect(&threadA, SIGNAL(started()), this, SLOT(onthreadA_started()));
    QObject::connect(&threadA, SIGNAL(finished()), this, SLOT(onthreadA_finished()));
    QObject::connect(&threadA, SIGNAL(newValue(int,int)), this, SLOT(onthreadA_newValue(int, int)));
    QObject::connect(ui->pushButton_enable, &QPushButton::clicked, this, &diceDialog::on_btnEnable);
    QObject::connect(ui->pushButton_start, &QPushButton::clicked, this, &diceDialog::on_btnStart);
    QObject::connect(ui->pushButton_pause, &QPushButton::clicked, this, &diceDialog::on_btnPause);
    QObject::connect(ui->pushButton_stop, &QPushButton::clicked, this, &diceDialog::on_btnStop);
    QObject::connect(ui->pushButton_clear, &QPushButton::clicked, this, &diceDialog::on_btnClear);
}
```

```cpp
diceDialog::~diceDialog()
{
    delete ui;
}
void diceDialog::onthreadA_started()
{
    ui->label->setText("thead status: thread started");
}
void diceDialog::onthreadA_finished()
{
    ui->label->setText("thead status: thread finished");
}
void diceDialog::onthreadA_newValue(int seq,int diceValue)
{
    ui->label->setText("thead status: newValue");
    QString str_temp=QString::asprintf("the  %d times play,points is: %d", seq, diceValue);
    ui->plainTextEdit->appendPlainText(str_temp);
    /* QImage image("../pics/1.jpg");
    ui->LabPic->setPixmap(QPixmap::fromImage(image));
    ui->LabPic->setGeometry(0,0,image.width(),image.height());*/
    /* QPixmap pic;
    QString filename=QString::asprintf("C:\\Users\\jun_s\\Desktop\\8chapter\\Qt_CMake_VS_Qthread-2\\pics\\%d.jpg", diceValue);
    pic.load(filename);
    ui->LabPic->setPixmap(pic); */

    QString filename=QString::asprintf("../pics/%d.jpg", diceValue);
    //QString filename("F:\\Study\\junior\\Qt\\door\\1.jpg");
    QImage *img=new QImage;
    if(!(img->load(filename)))  //加载图像
    {
        QMessageBox::information(this,
            tr(u8"打开图像失败"),
            tr(u8"打开图像失败!"));
        delete img;
        return;
    }
    ui->LabPic->resize(img->width(),img->height());
    ui->LabPic->setPixmap(QPixmap::fromImage(*img));

}
void diceDialog::closeEvent(QCloseEvent *event)
{
```

```cpp
        if(threadA.isRunning())
        {
            threadA.stopThread();
            threadA.wait();
        }
        event->accept();
}
/////////////////////////
void diceDialog::on_btnEnable()
{
    threadA.start();
    ui->pushButton_start->setEnabled(true);
    ui->pushButton_stop->setEnabled(false);
    ui->pushButton_pause->setEnabled(false);
    ui->pushButton_enable->setEnabled(false);

}
void diceDialog::on_btnStart()
{
    threadA.diceBegin();
    ui->pushButton_start->setEnabled(true);
    ui->pushButton_stop->setEnabled(true);
    ui->pushButton_pause->setEnabled(true);
    ui->pushButton_enable->setEnabled(false);
}
void diceDialog::on_btnPause()
{
    threadA.diceBegin();
    ui->pushButton_start->setEnabled(true);
    ui->pushButton_stop->setEnabled(false);
    ui->pushButton_pause->setEnabled(false);
    ui->pushButton_enable->setEnabled(false);
}
void diceDialog::on_btnStop()
{
    threadA.stopThread();
    threadA.wait();
    ui->pushButton_start->setEnabled(true);
    ui->pushButton_stop->setEnabled(false);
    ui->pushButton_pause->setEnabled(false);
    ui->pushButton_enable->setEnabled(false);
}
void diceDialog::on_btnClear()
```

```
{
    ui->plainTextEdit->clear();
}
```

5.2 基于 QMutex 和 QMutexLocker 多线程

本实验将在5.1实验的基础上进行改进,采用 QMutex 和 QMutexLocker 来进行线程的同步。实验的预期效果如图 5-2 所示。

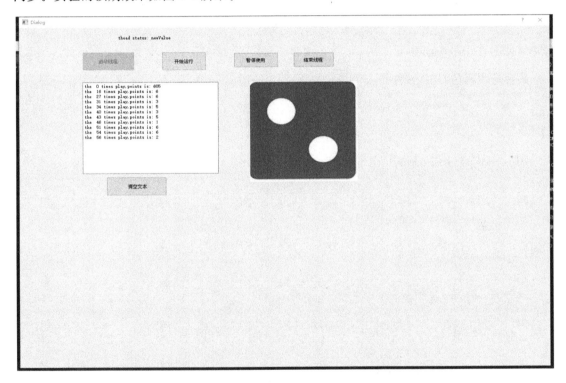

图 5-2 基于 QMutex 和 QMutexLocker 多线程的设计目标

首先是线程的 H 文件。

```
#ifndef QDICETHREAD_H
#define QDICETHREAD_H
#include<QObject>
#include<Qthread>
#include<QMutex>
class qdicethread: public QThread
{
    Q_OBJECT
private:
```

```cpp
        QMutex mutex;
        int m_seq = 0;
            int m_diceValue;
        bool m_Paused = true;
        bool m_stop = false;
protected:
        void run() Q_DECL_OVERRIDE;
public:
        qdicethread();
        void diceBegin();
        void dicePause();
        void stopThread();
        bool readValue(int &seq, int &diceValue);
};
#endif // QDICETHREAD_H
```

然后是线程的 CPP 文件。

```cpp
#include "qdicethread.h"
#include <Qtime>

qdicethread::qdicethread()
{
}
void qdicethread::diceBegin()
{
   m_Paused = false;
}
void qdicethread::dicePause()
{
   m_Paused = true;
}
void qdicethread::stopThread()
{
   m_stop = true;
}
void qdicethread::run()
{
   m_stop = false;
   m_seq = 0;
   QTime * time = newQTime();
   QString strTime;
   qsrand(time->currentTime().msec());
   while(!m_stop)
```

```
    {
        if(!m_Paused)
        {
            mutex.lock();
            m_diceValue = qrand();
            m_diceValue = (m_diceValue % 6) + 1;
            m_seq++;
            mutex.unlock();
        }
        msleep(500);
    }
    //quit();
}
bool qdicethread::readValue(int &seq, int &diceValue)
{
    if(mutex.tryLock())
    {
        seq = m_seq;
        diceValue = m_diceValue;
        mutex.unlock();
        return true;
    }
    else
    {
        return false;
    }
}
```

接着是对话框的 H 文件。

```
#ifndef DICEDIALOG_H
#define DICEDIALOG_H

#include <QDialog>
#include "qdicethread.h"
namespace Ui {
class diceDialog;
}
class diceDialog: public QDialog
{
    Q_OBJECT
private:
    qdicethread threadA;
protected:
```

```cpp
        void closeEvent(QCloseEvent * event);
public:
        explicit diceDialog(QWidget * parent = nullptr);
        ~diceDialog();
private slots:
        void onthreadA_started();
        void onthreadA_finished();

        void on_btnEnable();
        void on_btnStart();
        void on_btnPause();
        void on_btnStop();
        void on_btnClear();
private:
        Ui::diceDialog * ui;
        void canReadValue();
};
#endif // DICEDIALOG_H
```

最后是 CPP 文件。

```cpp
#include "diceDialog.h"
#include "ui_diceDialog.h"
#include <QImage>
#include <QDir>
#include <QLibrary>
#include <QString>
#include <QMessageBox>

diceDialog::diceDialog(QWidget * parent):
    QDialog(parent),
    ui(new Ui::diceDialog)
{
    ui->setupUi(this);
    QObject::connect(&threadA, SIGNAL(started()), this, SLOT(onthreadA_started()));
    QObject::connect(&threadA, SIGNAL(finished()), this, SLOT(onthreadA_finished()));
    QObject::connect(&threadA, SIGNAL(newValue(int,int)), this, SLOT(onthreadA_newValue(int, int)));
    QObject::connect(ui->pushButton_enable, &QPushButton::clicked, this, &diceDialog::on_btnEnable);
    QObject::connect(ui->pushButton_start, &QPushButton::clicked, this, &diceDialog::on_btnStart);
    QObject::connect(ui->pushButton_pause, &QPushButton::clicked, this, &diceDialog::on_btnPause);
```

```cpp
    QObject::connect(ui->pushButton_stop, &QPushButton::clicked, this, &diceDialog::on_
    btnStop);
    QObject::connect(ui->pushButton_clear, &QPushButton::clicked, this, &diceDialog::on_
    btnClear);
}
diceDialog::~diceDialog()
{
    delete ui;
}
void diceDialog::onthreadA_started()
{
    ui->label->setText("thead status: thread started");
}
void diceDialog::onthreadA_finished()
{
    ui->label->setText("thead status: thread finished");
}
void diceDialog::canReadValue()
{
    int  seq_temp = 0;
    int  diceValue_temp = 0;
    threadA.readValue(seq_temp, diceValue_temp);
    ui->label->setText("thead status: newValue");
    QString str_temp = QString::asprintf("the  %d times play,points is: %d", seq_temp,
    diceValue_temp);
    ui->plainTextEdit->appendPlainText(str_temp);
    /* QImage image("../pics/1.jpg");
    ui->LabPic->setPixmap(QPixmap::fromImage(image));
    ui->LabPic->setGeometry(0, 0, image.width(), image.height()); */
    /* QPixmap pic;
    QString filename = QString::asprintf("C:\\Users\\jun_s\\Desktop\\8chapter\\Qt_CMake_VS_
    Qthread-2\\pics\\%d.jpg", diceValue);
    pic.load(filename);
    ui->LabPic->setPixmap(pic); */

    QString filename = QString::asprintf("../pics/%d.jpg", diceValue_temp);
    //QString filename("F:\\Study\\junior\\Qt\\door\\1.jpg");
    QImage *img = new QImage;
    if(!(img->load(filename))) //加载图像
    {
        QMessageBox::information(this,
            tr(u8"打开图像失败"),
            tr(u8"打开图像失败!"));
```

```cpp
            delete img;
            return;
        }
        ui->LabPic->resize(img->width(), img->height());
        ui->LabPic->setPixmap(QPixmap::fromImage(*img));

}
void diceDialog::closeEvent(QCloseEvent *event)
{
    if(threadA.isRunning())
    {
        threadA.stopThread();
        threadA.wait();
    }
    event->accept();
}
////////////////////////////
void diceDialog::on_btnEnable()
{
    threadA.start();
    ui->pushButton_start->setEnabled(true);
    ui->pushButton_stop->setEnabled(false);
    ui->pushButton_pause->setEnabled(false);
    ui->pushButton_enable->setEnabled(false);


}
void diceDialog::on_btnStart()
{
    threadA.diceBegin();
    ui->pushButton_start->setEnabled(true);
    ui->pushButton_stop->setEnabled(true);
    ui->pushButton_pause->setEnabled(true);
    ui->pushButton_enable->setEnabled(false);

    canReadValue();
}
void diceDialog::on_btnPause()
{
    threadA.diceBegin();
    ui->pushButton_start->setEnabled(true);
    ui->pushButton_stop->setEnabled(false);
    ui->pushButton_pause->setEnabled(false);
    ui->pushButton_enable->setEnabled(false);
```

```
}
void diceDialog::on_btnStop()
{
    threadA.stopThread();
    threadA.wait();
    ui->pushButton_start->setEnabled(true);
    ui->pushButton_stop->setEnabled(false);
    ui->pushButton_pause->setEnabled(false);
    ui->pushButton_enable->setEnabled(false);
}
void diceDialog::on_btnClear()
{
    ui->plainTextEdit->clear();
}
```

采用了互斥机制后,在程序读写时,有效地避免了数字和图片显示不一致的情况的发生。

5.3 基于 Ontimer 显示多线程

采用 Ontimer 显示多线程方法后,预期的效果如图 5-3 所示。

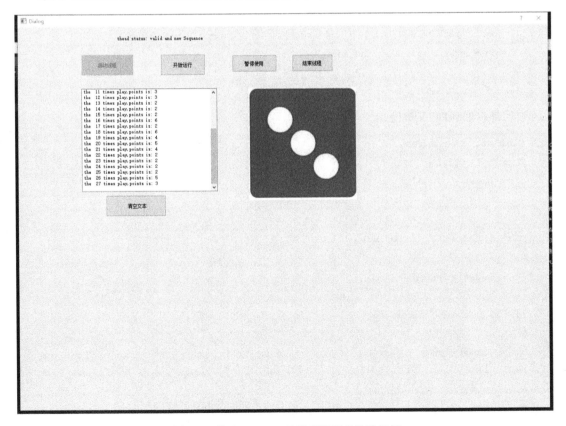

图 5-3 基于 Ontimer 显示多线程的设计目标

首先是多线程的 H 文件。

```cpp
#ifndef QDICETHREAD_H
#define QDICETHREAD_H

#include<QObject>
#include<Qthread>
#include<QMutex>
class qdicethread: public QThread
{
    Q_OBJECT
private:
    QMutex mutex;
    int m_seq = 0;
    int m_diceValue;
    bool m_Paused = true;
    bool m_stop = false;
protected:
    void run() Q_DECL_OVERRIDE;
public:
    qdicethread();
    void diceBegin();
    void dicePause();
    void stopThread();
    bool readValue(int &seq, int &diceValue);
};
#endif // QDICETHREAD_H
```

然后是多线程的 CPP 设计。

```cpp
#include "qdicethread.h"
#include<Qtime>
qdicethread::qdicethread()
{
}
void qdicethread::diceBegin()
{
    m_Paused = false;
}
void qdicethread::dicePause()
{
    m_Paused = true;
}
void qdicethread::stopThread()
{
```

```
        m_stop = true;
}
void qdicethread :: run()
{
    m_stop = false;
    m_seq = 0;
    QTime * time = newQTime();
    QString strTime;
    qsrand(time -> currentTime().msec());
    while(! m_stop)
    {
        if(! m_Paused)
        {
            QMutexLocker locker(&mutex);
            m_diceValue = qrand();
            m_diceValue = (m_diceValue % 6) + 1;
            m_seq ++ ;

        }
        msleep(500);
    }
    //quit();
}
bool qdicethread :: readValue(int &seq, int &diceValue)
{
    if(mutex.tryLock())
    {
        seq = m_seq;
        diceValue = m_diceValue;
        mutex.unlock();
        return true;
    }
    else
    {
        return false;
    }
}
```

接着是对话框的 H 文件。

```
# ifndef DICEDIALOG_H
# define DICEDIALOG_H


# include < QDialog >
```

```cpp
#include <QTime>
#include <QTimer>
#include "qdicethread.h"
namespace Ui {
class diceDialog;
}
class diceDialog: public QDialog
{
    Q_OBJECT
private:
    int mSeq, mDiceValue;
    qdicethread threadA;
    QTimer mTimer;
protected:
    void closeEvent(QCloseEvent * event);
public:
    explicit diceDialog(QWidget * parent = nullptr);
    ~diceDialog();
private slots:
    void onthreadA_started();
    void onthreadA_finished();
    void onTimeOut();

    void on_btnEnable();
    void on_btnStart();
    void on_btnPause();
    void on_btnStop();
    void on_btnClear();
private:
    Ui::diceDialog * ui;

};
#endif // DICEDIALOG_H
```

最后是对话框的 CPP 设计。

```cpp
#include "diceDialog.h"
#include "ui_diceDialog.h"
#include <QImage>
#include <QDir>
#include <QLibrary>
#include <QString>
#include <QMessageBox>

diceDialog::diceDialog(QWidget * parent);
```

```cpp
    QDialog(parent),
    ui(new Ui::diceDialog)
{
    ui->setupUi(this);
    QObject::connect(&threadA, SIGNAL(started()), this, SLOT(onthreadA_started()));
    QObject::connect(&threadA, SIGNAL(finished()), this, SLOT(onthreadA_finished()));
    QObject::connect(&mTimer, SIGNAL(timeout()), this, SLOT(onTimeOut()));

    QObject::connect(ui->pushButton_enable, &QPushButton::clicked, this, &diceDialog::on_btnEnable);
    QObject::connect(ui->pushButton_start, &QPushButton::clicked, this, &diceDialog::on_btnStart);
    QObject::connect(ui->pushButton_pause, &QPushButton::clicked, this, &diceDialog::on_btnPause);
    QObject::connect(ui->pushButton_stop, &QPushButton::clicked, this, &diceDialog::on_btnStop);
    QObject::connect(ui->pushButton_clear, &QPushButton::clicked, this, &diceDialog::on_btnClear);
}
diceDialog::~diceDialog()
{
    delete ui;
}
void diceDialog::onthreadA_started()
{
    ui->label->setText("thead status: thread started");
}
void diceDialog::onthreadA_finished()
{
    ui->label->setText("thead status: thread finished");
}
void diceDialog::onTimeOut()
{
    int  seq_temp = 0;
    int  diceValue_temp = 0;
    bool valid = threadA.readValue(seq_temp, diceValue_temp);
    if(valid &&(seq_temp != mSeq))
    {
        ui->label->setText("thead status: valid and new Sequence");
        QString str_temp = QString::asprintf("the  %d times play,points is: %d", seq_temp, diceValue_temp);
        ui->plainTextEdit->appendPlainText(str_temp);
```

```cpp
        QString filename = QString::asprintf("C:\\Users\\CTOS\\Desktop\\8chapter\\8chapter
            \\Qt_CMake_VS_Qthread-2\\pics\\%d.jpg", diceValue_temp);
        //QString filename("F:\\Study\\junior\\Qt\\door\\1.jpg");
        QImage *img = new QImage;
        if(!(img->load(filename)))  //加载图像
        {
            QMessageBox::information(this,
                tr(u8"打开图像失败"),
                tr(u8"打开图像失败!"));
            delete img;
            return;
        }
        ui->LabPic->resize(img->width(), img->height());
        ui->LabPic->setPixmap(QPixmap::fromImage(*img));
    }
    else
    {
        ui->label->setText("thead status: novalid or no new Sequence");
    }

}
void diceDialog::closeEvent(QCloseEvent *event)
{
    if(threadA.isRunning())
    {
        threadA.stopThread();
        threadA.wait();
    }
    event->accept();
}
///////////////////////////
void diceDialog::on_btnEnable()
{
    mSeq = 0;
    threadA.start();
    ui->pushButton_start->setEnabled(true);
    ui->pushButton_stop->setEnabled(false);
    ui->pushButton_pause->setEnabled(false);
    ui->pushButton_enable->setEnabled(false);

}
```

```cpp
void diceDialog::on_btnStart()
{
    threadA.diceBegin();
    mTimer.start(500);
    ui->pushButton_start->setEnabled(true);
    ui->pushButton_stop->setEnabled(true);
    ui->pushButton_pause->setEnabled(true);
    ui->pushButton_enable->setEnabled(false);
}
void diceDialog::on_btnPause()
{
    threadA.dicePause();
    ui->pushButton_start->setEnabled(true);
    ui->pushButton_stop->setEnabled(false);
    ui->pushButton_pause->setEnabled(false);
    ui->pushButton_enable->setEnabled(false);
}
void diceDialog::on_btnStop()
{

    threadA.stopThread();
    threadA.wait();
    ui->pushButton_start->setEnabled(true);
    ui->pushButton_stop->setEnabled(false);
    ui->pushButton_pause->setEnabled(false);
    ui->pushButton_enable->setEnabled(false);


}
void diceDialog::on_btnClear()
{
    ui->plainTextEdit->clear();


}
```

采用 OnTime 方法的主要区别在于用定时器的信号和槽的机制替代了原本的带参数传递的信号和槽的机制。

5.4 基于 Producer 与 Consumer 显示多线程

采用 Producer 与 Consumer 显示多线程方法后，预期的效果如图 5-4 所示。

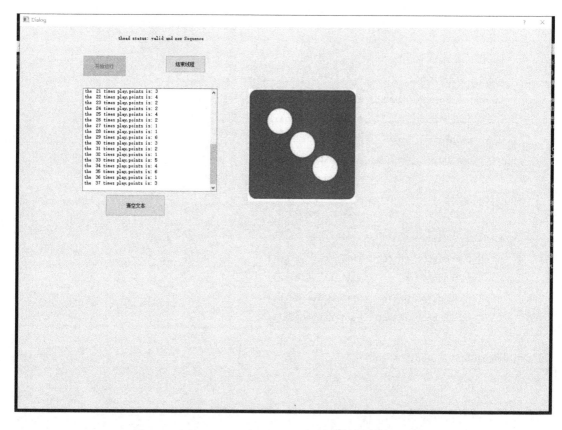

图 5-4 基于 Producer 与 Consumer 显示多线程的设计目标

首先是多线程的 H 文件。

```
#ifndef QDICETHREAD_H
#define QDICETHREAD_H

#include<QObject>
#include<Qthread>
#include<Qtime>
#include<QMutex>
#include<QWaitCondition>
class QThreadProducer: public QThread
{
    Q_OBJECT
public:
    bool m_stop = false;
    QMutex mutex;
    QWaitCondition newdataAvailable;
    int seq;
    int diceValue;
protected:
```

```
        void run() Q_DECL_OVERRIDE;
public:
    QThreadProducer();
    void stopThread();
};
Class QThreadConsumer: public QThread
{
    Q_OBJECT
public:
    bool m_stop;
    QMutex mutex;
    QWaitCondition newdataAvailable;
    int seq ;
    int diceValue;
protected:
    void run() Q_DECL_OVERRIDE;
public:
    QThreadConsumer();
    void stopThread();
};
# endif // QDICETHREAD_H
```

然后是多线程的 CPP 文件。

```
# include "qdicethread.h"

QThreadProducer :: QThreadProducer()
{
}
void QThreadProducer :: run()
{
    m_stop = false;
    seq = 0;
    qsrand(QTime :: currentTime().msec());
    while(!m_stop)
    {
    mutex.lock();
        diceValue = qrand();
        diceValue = (diceValue % 6) + 1;
        seq ++ ;
        mutex.unlock();
        newdataAvailable.wakeAll();
        msleep(500);
    }
```

```cpp
}
//consumer
QThreadConsumer::QThreadConsumer()
{
}
void QThreadConsumer::run()
{
    m_stop = false;

    while(!m_stop)
    {
        mutex.lock();
        newdataAvailable.wait(&mutex);
        mutex.unlock();
    }
}
```

接着是对话框的 H 文件。

```cpp
#ifndef DICEDIALOG_H
#define DICEDIALOG_H

#include <QDialog>
#include <QTime>
#include <QTimer>
#include "qdicethread.h"
namespace Ui {
class diceDialog;
}
class diceDialog: public QDialog
{
    Q_OBJECT
private:
    int mSeq, mDiceValue;
    QThreadProducer threadA;
    QThreadConsumer threadB;
    QTimer mTimer;
protected:
    void closeEvent(QCloseEvent * event);
public:
    explicit diceDialog(QWidget * parent = nullptr);
    ~diceDialog();
private slots:
    void onthreadA_started();
```

```
    void onthreadA_finished();
    void onthreadB_started();
    void onthreadB_finished();
    void onTimeOut();
    void on_btnStart();
    void on_btnStop();
    void on_btnClear();
private:
    Ui::diceDialog *ui;

};
#endif // DICEDIALOG_H
```

最后是对话框的 CPP 文件。

```
#include "diceDialog.h"
#include "ui_diceDialog.h"
#include <QImage>
#include <QDir>
#include <QLibrary>
#include <QString>
#include <QMessageBox>

diceDialog::diceDialog(QWidget *parent) :
    QDialog(parent),
    ui(new Ui::diceDialog)
{
    ui->setupUi(this);
    QObject::connect(&threadA, SIGNAL(started()), this, SLOT(onthreadA_started()));
    QObject::connect(&threadA, SIGNAL(finished()), this, SLOT(onthreadA_finished()));
    QObject::connect(&threadB, SIGNAL(started()), this, SLOT(onthreadB_started()));
    QObject::connect(&threadB, SIGNAL(finished()), this, SLOT(onthreadB_finished()));
    QObject::connect(&mTimer, SIGNAL(timeout()), this, SLOT(onTimeOut()));
    QObject::connect(ui->pushButton_start, &QPushButton::clicked, this, &diceDialog::on_btnStart);
    QObject::connect(ui->pushButton_stop, &QPushButton::clicked, this, &diceDialog::on_btnStop);
    QObject::connect(ui->pushButton_clear, &QPushButton::clicked, this, &diceDialog::on_btnClear);
}
diceDialog::~diceDialog()
{
    delete ui;
}
```

```cpp
void diceDialog::onthreadA_started()
{
    ui->label->setText("thead status: threadA started");
}
void diceDialog::onthreadA_finished()
{
    ui->label->setText("thead status: threadA finished");
}
void diceDialog::onthreadB_started()
{
    ui->label->setText("thead status: threadB started");
}
void diceDialog::onthreadB_finished()
{
    ui->label->setText("thead status: threadB finished");
}
void diceDialog::onTimeOut()
{
    int  seq_temp = 0;
    int  diceValue_temp = 0;
    threadA.start();
    threadB.start();
    seq_temp = threadA.seq;
    diceValue_temp = threadA.diceValue;
    ui->label->setText("thead status: valid and new Sequence");
    QString str_temp = QString::asprintf("the  %d times play,points is: %d", seq_temp, diceValue_temp);
    ui->plainTextEdit->appendPlainText(str_temp);
    QString filename = QString::asprintf("C:\\Users\\CTOS\\Desktop\\8chapter\\8chapter\\Qt_CMake_VS_Qthread-2\\pics\\%d.jpg", diceValue_temp);
    //QString filename("F:\\Study\\junior\\Qt\\door\\1.jpg");
    QImage *img = new QImage;
    if(!(img->load(filename)))  //加载图像
    {
        QMessageBox::information(this,
            tr(u8"打开图像失败"),
            tr(u8"打开图像失败!"));
        delete img;
        return;
    }
    ui->LabPic->resize(img->width(), img->height());
    ui->LabPic->setPixmap(QPixmap::fromImage(*img));

}
```

```cpp
void diceDialog::closeEvent(QCloseEvent * event)
{
    if(threadA.isRunning())
    {
        threadA.terminate();
        threadA.wait();
    }
    if(threadB.isRunning())
    {
        threadB.terminate();
        threadB.wait();
    }
    event->accept();
}
/////////////////////////
void diceDialog::on_btnStart()
{
    threadA.start();
    threadB.start();
    mTimer.start(500);
    ui->pushButton_start->setEnabled(false);
    ui->pushButton_stop->setEnabled(true);

}
void diceDialog::on_btnStop()
{
    threadA.terminate();
    threadA.wait();
    threadB.terminate();
    threadB.wait();
    ui->pushButton_start->setEnabled(true);
    ui->pushButton_stop->setEnabled(false);

}
void diceDialog::on_btnClear()
{

    ui->plainTextEdit->clear();

}
```

本章通过建立基本的线程,采用互斥的方式和定时器的方法,以及采用生产者-消费者模式,将线程的应用进行了详细讲解。

思 考 题

简答题
1. QT 实现多线程的方法有哪些？
2. 如何采用多线程实现多相机的图像采集？
3. 如何采用多线程实现板卡的运动控制？

第6章 QT、OpenCV、HIK 图像采集系统

本章介绍采用两个网络工业相机和海康工业相机的 SDK 来驱动网络相机的例子。本实验采用 VS 2019，QT 5.15.0，Win 10，OpenCV 4.5.3，通过 C++ 的驱动进行双相机的图像采集。本次实验是基于软触发的模式完成图像的采集。通过本次实验，我们可以掌握 C++ 的类的构造方法、虚函数的使用和工业相机的二次开发。其中，6.1 节介绍基本类的构造过程和核心函数；6.2 节重点介绍虚函数在接口函数定义的应用，其目的是让接口函数统一的同时，不同的相机分别采用不同的 SDK 实现。

6.1 基于 Hik-QT-IMG 软触发采集图像

基于海康彩色工业相机的图像采集平台如图 6-1 所示。

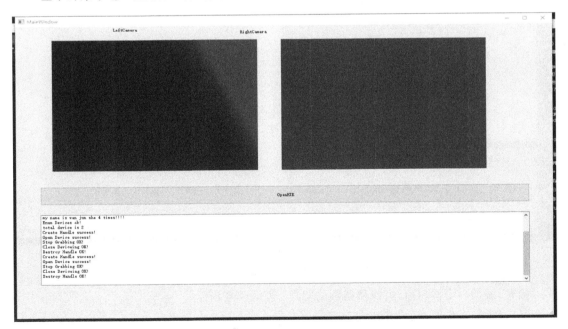

图 6-1 基于海康彩色工业相机的图像采集平台

C++语言与机器视觉编程实战

首先将海康相机的头文件引入,然后定义两个相机的指针。如图 6-2 所示。

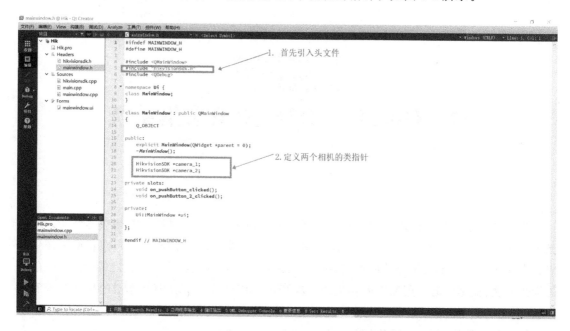

图 6-2 编程方法中头文件的引用和类指针

然后,在构造函数进行变量赋值和析构函数对对象的销毁。如图 6-3 所示。

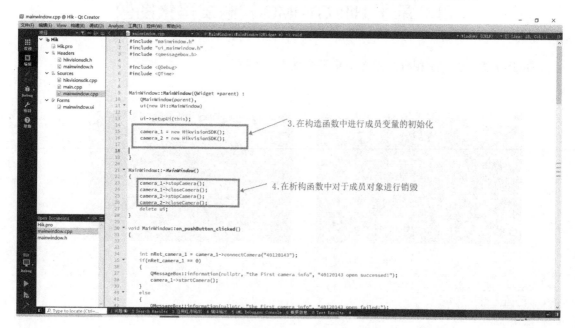

图 6-3 海康相机构造函数与析构函数

接着,添加一个按钮后,进行槽函数的实现。如图 6-4 所示。

最后,进行采集图像。如图 6-5 所示。

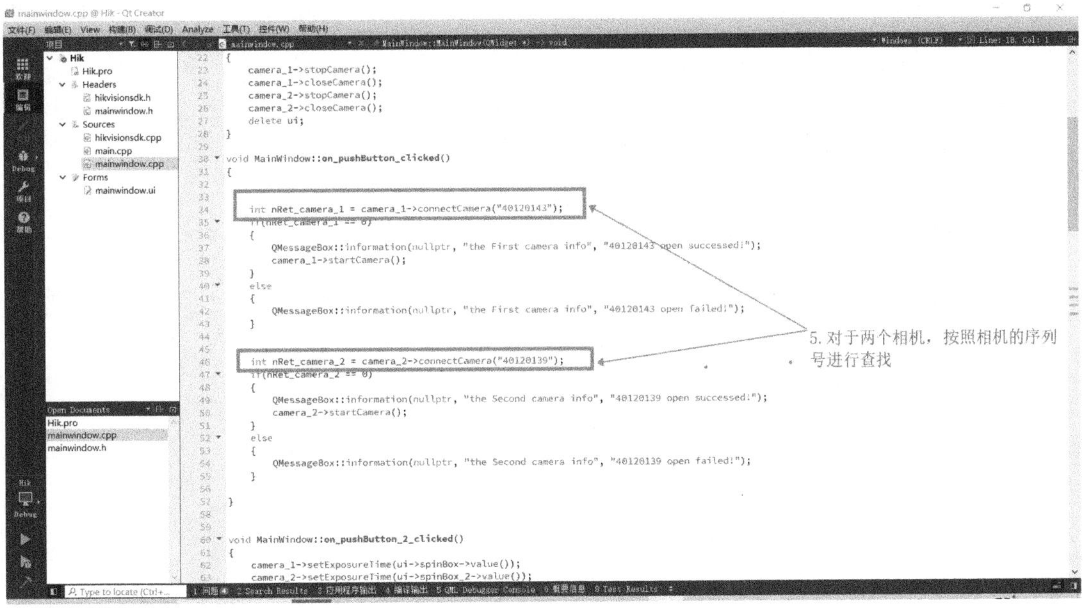

第 6 章 | QT、OpenCV、HIK 图像采集系统

图 6-4 海康相机序列号查找相机

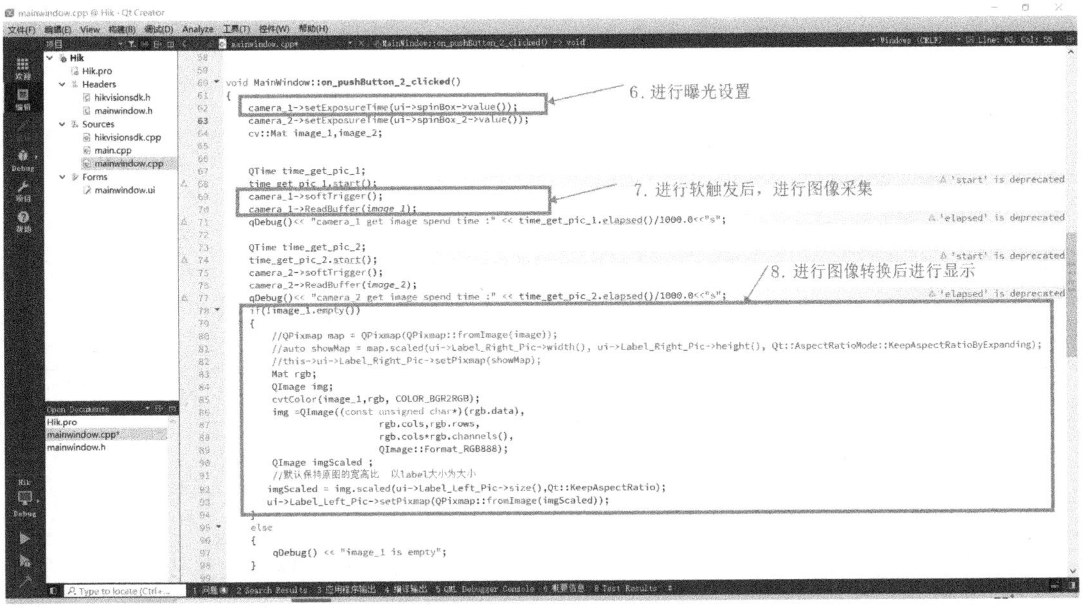

图 6-5 海康相机曝光值设置、软触发、图像转换后的显示

6.2 纯虚函数 Virtural 接口方法

采用纯虚函数是一个比较好的接口函数构建方式。接下来,我们就采用纯虚函数来构建海康相机的接口函数。

首先是 camerainterface.h 通用相机的头文件,如下所示。

```cpp
#ifndef CAMERAINTERFACE_H
#define CAMERAINTERFACE_H

#include <exception>
#include <string>
#include <filesystem>
#include <map>
#include <QObject>
#include <QVariant>
#include "opencv2/opencv.hpp"
class CameraError: public std::exception
{
public:
    CameraError(const std::string& w):msg(w) {};
    ~CameraError() {};
    const char * what() const { return msg.data(); };
private:
    std::string msg;
};

enum CameraTriggerType
{
    SoftTrigger,
    HardTrigger,
    AutoTrigger
};

enum CameraFlip
{
    DoNotFlip,
    RoundX,
    RoundY,
    Both
};

enum CameraTransPose
{
    DoNotTransPose,
    Rotate90
};

namespace cv
```

```cpp
{
    class Mat;
}
class QImage;
class CameraInterface: public QObject
{
    Q_OBJECT
public:
    CameraInterface(const std::string& taskName = "null"): type(CameraTriggerType::
    AutoTrigger), taskName(taskName) {};
    virtual ~CameraInterface() {}
    virtual bool isOpen() = 0;
    virtual void openCamera(const int&) = 0;
    virtual void openWithSn(const std::string&) = 0;
    virtual void setCameraTriggerType(const CameraTriggerType& t)
    {
        type = t;
    }
    CameraTriggerType getCameraTriggerType()
    {
        return type;
    }
    virtual cv::Mat getCvMat(const int&, const bool& flag = false) = 0;
    virtual QImage getQImage(const int&, const bool& flag = false) = 0;
    virtual std::map<std::string, int> getCameraList() = 0;
    virtual std::string getSerialName() = 0;
    virtual void showCameraSettingDialog() = 0;
    virtual void changeParamFromMap(const std::map<std::string, std::tuple<std::
    string, std::string>>&) = 0;
    virtual void getParamFromCamera(std::map<std::string, std::tuple<std::string, std::
    string>>&) = 0;
    virtual std::string getTaskName() { return taskName; }
    virtual void startGrabImage(const int& delay) = 0;
    void setFlip(const CameraFlip& fp)
    {
        flip = fp;
    }
    void setTranspose(const CameraTransPose& rt)
    {
        transpose = rt;
    }
    CameraFlip getFlip()
    {
```

```cpp
            return flip;
        }
        CameraTransPose getTranspose()
        {
            return transpose;
        }
    signals:
        void grabedImage(QVariant img);
    protected:
        std::string taskName;
        CameraFlip flip;
        CameraTransPose transpose;
    private:
        CameraTriggerType type;
    };
    Q_DECLARE_METATYPE(cv::Mat)

    #endif // CAMERAINTERFACE_H
```

然后是海康相机的头文件,如下所示。

```cpp
#pragma once
#include <QObject>
#include <QImage>
#include "cameraInterface.h"
class HikCameraInterface : public CameraInterface
{
    Q_OBJECT
public:
    HikCameraInterface(const std::string& taskName = "null", QObject * parent = nullptr);
    ~HikCameraInterface();
public slots:
    virtual bool isOpen() override;
    virtual void openCamera(const int&) override;
    virtual void openWithSn(const std::string&) override;
    virtual cv::Mat getCvMat(const int&, const bool& flag = false) override;
    virtual QImage getQImage(const int&, const bool& flag = false) override;
    virtual std::map<std::string, int> getCameraList() override;
    virtual std::string getSerialName() override;
    virtual void showCameraSettingDialog() override;
    virtual void changeParamFromMap(const std::map<std::string, std::tuple<std::string, std::string>> &) override;
    virtual void getParamFromCamera(std::map<std::string, std::tuple<std::string, std::string>> &) override;
```

```cpp
    virtual void startGrabImage(const int& delay) override;
private:
    bool opened;
    void * handle;
    std :: mutex mutex;
    int type;
    unsigned int g_nPayloadSize;
    std :: string serial_number;
    std :: mutex param_mutex;
    cv :: Mat imgBuffer;
};
```

接着是实现这些纯虚函数的 CPP 文件:HikCameraInterface.cpp,如下所示。

```cpp
#include "HikCameraInterface.h"
#include "MvCameraControl.h"
#include "opencv2/opencv.hpp"
#include <QtConcurrent>
#include <qdebug.h>
HikCameraInterface :: HikCameraInterface(const std :: string& taskName, QObject * parent)
    :CameraInterface(taskName),
    opened(false)
{
}

HikCameraInterface :: ~HikCameraInterface()
{
}

void HikCameraInterface :: openCamera(const int&)
{
}

void HikCameraInterface :: openWithSn(const std :: string& serial)
{
    uint nRet = MV_OK;
    MV_CC_DEVICE_INFO_LIST stDeviceList;
    memset(&stDeviceList, 0, sizeof(MV_CC_DEVICE_INFO_LIST));
    nRet = MV_CC_EnumDevices(MV_GIGE_DEVICE | MV_USB_DEVICE, &stDeviceList);
    if(MV_OK != nRet)
    {
        throw CameraError("MV_CC_EnumDevices error," + std :: to_string(nRet));
    }

    if(stDeviceList.nDeviceNum <= 0)
```

```cpp
    {
        throw CameraError(u8"未找到相机！\n");
    }
    unsigned int idx = 0;
    bool flag = false;
    for(unsigned int i = 0; i < stDeviceList.nDeviceNum; i++)
    {
        std::string tmp;
        switch(stDeviceList.pDeviceInfo[i]->nTLayerType)
        {
        case MV_USB_DEVICE:
            for(int j = 0; j < 64; j++)
            {
                if(stDeviceList.pDeviceInfo[i]->SpecialInfo.stUsb3VInfo.chSerialNumber[j] == '\0')
                    break;
                tmp.push_back(char(stDeviceList.pDeviceInfo[i]->SpecialInfo.stUsb3VInfo.chSerialNumber[j]));
            }
            break;
        case MV_GIGE_DEVICE:
            for(int j = 0; j < 64; j++)
            {
                if(stDeviceList.pDeviceInfo[i]->SpecialInfo.stGigEInfo.chSerialNumber[j] == '\0')
                    break;
                tmp.push_back(char(stDeviceList.pDeviceInfo[i]->SpecialInfo.stGigEInfo.chSerialNumber[j]));
            }
            break;
        case MV_CAMERALINK_DEVICE:
            for(int j = 0; j < 64; j++)
            {
                if(stDeviceList.pDeviceInfo[i]->SpecialInfo.stCamLInfo.chSerialNumber[j] == '\0')
                    break;
                tmp.push_back(char(stDeviceList.pDeviceInfo[i]->SpecialInfo.stCamLInfo.chSerialNumber[j]));
            }
            break;
        default:
            break;
```

```cpp
        }
        if(tmp == serial)
        {
            flag = true;
            idx = i;
            break;
        }
    }
    if(!flag)
    {
        throw CameraError(u8"打开相机失败:" + serial + "\n");
    }
    if(idx >= stDeviceList.nDeviceNum)
    {
        throw CameraError(u8"打开相机失败 index:\n" + std::to_string(idx));
    }
    nRet = MV_CC_CreateHandle(&handle, stDeviceList.pDeviceInfo[idx]);
    if(MV_OK != nRet)
    {
        throw CameraError(u8"打开相机失败!" + std::to_string(nRet) + "\n");
    }
    nRet = MV_CC_OpenDevice(handle);
    if(MV_OK != nRet)
    {
        throw CameraError(u8"打开相机失败!" + std::to_string(nRet) + "\n");
    }
    MV_CC_SetAcquisitionMode(handle, 2);
    if(MV_OK != nRet)
    {
        throw CameraError(u8"设置相机采集模式失败!" + std::to_string(nRet) + "\n");
    }
    nRet = MV_CC_SetEnumValueByString(handle, "TriggerMode", "On");
    if(MV_OK != nRet)
    {
        throw CameraError(u8"设置触发模式失败!" + std::to_string(nRet) + "\n");
    }

    nRet = MV_CC_SetBoolValue(handle, "GammaEnable", false);
    if(MV_OK != nRet)
    {
        throw CameraError(u8"设置 GammaEnable 失败!" + std::to_string(nRet) + "\n");
    }
    nRet = MV_CC_SetEnumValueByString(handle, "TriggerSource", "Software");
```

```cpp
        if(MV_OK != nRet)
        {
            throw CameraError(u8"设置触发源失败!" + std :: to_string(nRet) + "\n");
        }
        //nRet = MV_CC_SetTriggerSource(handle, MV_TRIGGER_SOURCE_SOFTWARE);
        nRet = MV_CC_StartGrabbing(handle);
        if(MV_OK != nRet)
        {
            throw CameraError(u8"开启取图使能失败!" + std :: to_string(nRet) + "\n");
        }

        MVCC_INTVALUE stParam;
        memset(&stParam, 0, sizeof(MVCC_INTVALUE));
        nRet = MV_CC_GetIntValue(handle, "PayloadSize", &stParam);
        if(MV_OK != nRet)
        {
            throw CameraError("Get PayloadSize fail! nRet " + std :: to_string(nRet) + "\n");
        }

        nRet = MV_CC_SetEnumValueByString(handle, "ExposureAuto", "Off");
        if(MV_OK != nRet)
        {
            throw CameraError(u8"曝光模式设置失败! nRet " + std :: to_string(nRet) + "\n");
        }
        nRet = MV_CC_SetEnumValueByString(handle, "ExposureAuto", "Off");
        if(MV_OK != nRet)
        {
            throw CameraError(u8"增益模式设置失败! nRet " + std :: to_string(nRet) + "\n");
        }
        serial_number = serial;
        g_nPayloadSize = stParam.nCurValue;
        opened = true;
}

cv :: Mat HikCameraInterface :: getCvMat(const int& cn, const bool& flag)
{
    std :: lock_guard< std :: mutex > locker(mutex);
    auto test_start = std :: chrono :: system_clock :: now();
    //return cv :: imread("./localImg.bmp");
    /* cv :: Mat imgGrey;
    if(img.channels() == 3)
        cv :: cvtColor(img, imgGrey, cv :: COLOR_BGR2GRAY);
    return imgGrey; */
```

```cpp
        if(!isOpen())
        {
            //return cv::imread("testCam.bmp");
#ifdef _DEBUG
            return cv::imread("D://PF32RQ9S(1).bmp");
#endif // DEBUG
            throw CameraError("Camera is closed.\n");
        }
        uint nRet = MV_OK;
        cv::Mat out;
        if(!opened)
            return out;
        MV_FRAME_OUT_INFO_EX stImageInfo = {0};
        memset(&stImageInfo, 0, sizeof(MV_FRAME_OUT_INFO_EX));
        unsigned char * pData = (unsigned char *)malloc(sizeof(unsigned char) * (g_nPayloadSize));
        if(pData == NULL)
        {
            throw CameraError("Allocate memory failed.\n");
            return out;
        }

        auto t = std::chrono::duration_cast<std::chrono::milliseconds>(std::chrono::system_clock::now()-test_start).count();
        qDebug()<<t;
        // get one frame from camera with timeout = 1000ms
        nRet = MV_CC_TriggerSoftwareExecute(handle);
        if(nRet != MV_OK)
        {
            free(pData);
            pData = NULL;
            throw CameraError("MV_CC_TriggerSoftwareExecute_error" + std::to_string(nRet) + "\n");
            return out;
        }
        nRet = MV_CC_GetOneFrameTimeout(handle, pData, g_nPayloadSize, &stImageInfo, 8000);
        if(nRet != MV_OK)
        {
            free(pData);
            pData = NULL;
            throw CameraError("No data" + std::to_string(nRet) + "\n");
            return out;
        }
```

```cpp
auto t2 = std::chrono::duration_cast<std::chrono::milliseconds>(std::chrono::system_clock::now()-test_start).count();
qDebug()<< t2;
MV_CC_PIXEL_CONVERT_PARAM param;
param.nHeight = stImageInfo.nHeight;
param.nWidth = stImageInfo.nWidth;
param.pSrcData = pData;
param.enSrcPixelType = stImageInfo.enPixelType;
param.enDstPixelType = PixelType_Gvsp_RGB8_Packed;
//param.enDstPixelType = PixelType_Gvsp_Mono8;
param.nSrcDataLen = stImageInfo.nFrameLen;
int bufSize = stImageInfo.nHeight * stImageInfo.nWidth * 3;
unsigned char * cData = (unsigned char *)malloc(sizeof(unsigned char) * (bufSize));
param.pDstBuffer = cData;
param.nDstBufferSize = bufSize;
nRet = MV_CC_ConvertPixelType(handle, &param);

if(nRet != MV_OK)
{
    free(pData);
    free(cData);
    throw CameraError("MV_CC_ConvertPixelType err " + std::to_string(nRet) + "\n");
    return out;
}

//设置相机三通道白平衡
/* MV_CC_SetBalanceRatioRed(handle,136);
MV_CC_SetBalanceRatioGreen(handle, 100);
MV_CC_SetBalanceRatioBlue(handle, 64); */

auto t3 = std::chrono::duration_cast<std::chrono::milliseconds>(std::chrono::system_clock::now()-test_start).count();
qDebug()<< t3;
cv::Mat srcImage = cv::Mat(stImageInfo.nHeight, stImageInfo.nWidth, CV_8UC3, cData);

/* if(param.enPixelType == PixelType_Gvsp_RGB8_Packed)
{
    srcImage = cv::Mat(stImageInfo.nHeight, stImageInfo.nWidth, CV_8UC3, cData);
}
else
{
    printf("unsupported pixel format -> %d\n", stImageInfo.enPixelType);
    return out;
```

```cpp
} */
switch(cn)
{
case 1:
    cv::cvtColor(srcImage, out, cv::COLOR_RGB2GRAY);
    break;
case 3:
    cv::cvtColor(srcImage, out, cv::COLOR_RGB2BGR);
    break;
default:
    break;
}
auto t4 = std::chrono::duration_cast<std::chrono::milliseconds>(std::chrono::system_clock::now()-test_start).count();
qDebug()<<t4;
free(pData);
free(cData);
switch(getTranspose())
{
case CameraTransPose::DoNotTransPose:
    break;
case CameraTransPose::Rotate90:
    cv::transpose(out, out);
    break;
default:
    break;
}

switch(getFlip())
{
case CameraFlip::DoNotFlip:
    break;
case CameraFlip::RoundX:
    cv::flip(out, out, 0);
    break;
case CameraFlip::RoundY:
    cv::flip(out, out, 1);
    break;
case CameraFlip::Both:
    cv::flip(out, out, -1);
    break;
default:
    break;
```

```cpp
    }
    auto t5 = std::chrono::duration_cast<std::chrono::milliseconds>(std::chrono::
system_clock::now()-test_start).count();
    qDebug()<< t5;
    qDebug()<<"done";
    return out;
}

QImage HikCameraInterface::getQImage(const int&, const bool& flag)
{
    return QImage();
}


std::map<std::string, int> HikCameraInterface::getCameraList()
{
    return std::map<std::string, int>();
}


std::string HikCameraInterface::getSerialName()
{
    return std::string();
}


void HikCameraInterface::showCameraSettingDialog()
{
}


void HikCameraInterface::changeParamFromMap(const std::map<std::string, std::tuple<std::
string, std::string>>& mp)
{
    std::lock_guard<std::mutex> locker(param_mutex);
    int nRet = MV_OK;
    for(auto& node: mp)
    {
        auto [type, val] = node.second;
        if(type == "float")
        {
            nRet = MV_CC_SetFloatValue(handle, node.first.c_str(), std::atof(val.c_str
())));
            if(nRet != MV_OK)
            {
                throw CameraError(node.first + " error:" + std::to_string(nRet) + "\n");
```

```cpp
        }
    }
    else if(type == "int")
    {
        nRet = MV_CC_SetIntValue(handle, node.first.c_str(), std::atoi(val.c_str()));
        if(nRet != MV_OK)
        {
            throw CameraError(node.first + " error:" + std::to_string(nRet) + "\n");
        }
    }
    else if(type == "enum")
    {
        nRet = MV_CC_SetEnumValue(handle, node.first.c_str(), std::atoi(val.c_str()));
        if(nRet != MV_OK)
        {
            throw CameraError(node.first + " error:" + std::to_string(nRet) + "\n");
        }
    }
    else if(type == "string")
    {
        nRet = MV_CC_SetStringValue(handle, node.first.c_str(), val.c_str());
        if(nRet != MV_OK)
        {
            throw CameraError(node.first + " error:" + std::to_string(nRet) + "\n");
        }
    }
    else if(type == "bool")
    {
        nRet = MV_CC_SetBoolValue(handle, node.first.c_str(), std::atoi(val.c_str()));
        if(nRet != MV_OK)
        {
            throw CameraError(node.first + " error:" + std::to_string(nRet) + "\n");
        }
    }
    }
}

void HikCameraInterface::getParamFromCamera(std::map<std::string, std::tuple<std::string, std::string>> & mp)
{
    std::lock_guard<std::mutex> locker(param_mutex);
    if(mp.empty())
    {
```

```cpp
            mp = std::map<std::string, std::tuple<std::string, std::string>>{
                {"ExposureTimeAbs",std::make_tuple("float","")}
                ,{"GainRaw",std::make_tuple("int","")}
            };
        }
        for(auto& node: mp)
        {
            auto& [type, val] = node.second;
            if(type == "float")
            {
                MVCC_FLOATVALUE rtn_val;
                MV_CC_GetFloatValue(handle, node.first.c_str(), &rtn_val);
                val = std::to_string(rtn_val.fCurValue);
            }
            else if(type == "int")
            {
                MVCC_INTVALUE rtn_val;
                MV_CC_GetIntValue(handle, node.first.c_str(), &rtn_val);
                val = std::to_string(rtn_val.nCurValue);
            }
            else if(type == "enum")
            {
                MVCC_ENUMVALUE rtn_val;
                MV_CC_GetEnumValue(handle, node.first.c_str(), &rtn_val);
                val = std::to_string(rtn_val.nCurValue);
            }
            else if(type == "string")
            {
                MVCC_STRINGVALUE rtn_val;
                MV_CC_GetStringValue(handle, node.first.c_str(), &rtn_val);
                val = rtn_val.chCurValue;
            }
            else if(type == "bool")
            {
                bool rtn_val;
                MV_CC_GetBoolValue(handle, node.first.c_str(), &rtn_val);
                val = std::to_string(rtn_val);
            }
        }
    }

bool HikCameraInterface::isOpen()
{
```

```cpp
    return opened;
}

void HikCameraInterface::startGrabImage(const int& delay)
{
    auto future = QtConcurrent::run([this, delay]() {
        QThread::msleep(delay);
        auto img = getCvMat(1);
        QVariant var;
        var.setValue(img);
        emit grabedImage(var);
    });
}
```

如果需要使用这个类,首先需要定义一个头文件,如下所示。

```cpp
std::shared_ptr<CameraInterface> camera_1;
std::shared_ptr<CameraInterface> camera_2;
    void writeParamToCamera(std::shared_ptr<CameraInterface> camPtr, const int& exp,
    const int& gain);
```

然后是调用方法,在构造函数中进行初始化,如下所示。

```cpp
camera_1 = std::make_shared<HikCameraInterface>();
camera_2 = std::make_shared<HikCameraInterface>();
```

接着,采用 SN 打开相机,如下所示。

```cpp
try {
        camera_1->openWithSn("40120143");
        QMessageBox::information(nullptr, "the first camera info", "40120143 open successed!");
        m_log->write("the first camera info 40120143 open successed!");
        ui->plainTextEdit->appendPlainText("the first camera info 40120143 open successed!");
    } catch(CameraError& ex) {
        QMessageBox::information(nullptr,"the first camera info",ex.what());
    }
```

再接着进行曝光时间设置,如下所示。

```cpp
void MainWindow::writeParamToCamera(std::shared_ptr<CameraInterface> camPtr, const int&
exp, const int& gain)
{
    std::map<std::string, std::tuple<std::string, std::string>> cparam;
    camPtr->getParamFromCamera(cparam);
    std::get<1>(cparam[exposure_node_name]) = std::to_string(exp);
    std::get<1>(cparam[gain_node_name]) = std::to_string(gain);
    camPtr->changeParamFromMap(cparam);
}
```

最后进行图像采集与显示,如下所示。

```cpp
writeParamToCamera(camera_1,ui->spinBox->value(),1);
    QThread::msleep(300);
    cv::Mat image_1,image_2;
    try{
        image_1 = camera_1->getCvMat(3);
    }catch(CameraError& ex){
        QMessageBox::information(nullptr,"info",ex.what());
    }
    if(!image_1.empty())
    {
        cv::Mat rgb;
        QImage img;
        cvtColor(image_1,rgb,cv::COLOR_BGR2RGB);
        img = QImage((const unsigned char *)(rgb.data),
                    rgb.cols,rgb.rows,
                    rgb.cols * rgb.channels(),
                    QImage::Format_RGB888);
        QImage imgScaled ;
        //默认保持原图的宽高比    以label大小为大小
        imgScaled = img.scaled(ui->Label_Left_Pic->size(),Qt::KeepAspectRatio);
        ui->Label_Left_Pic->setPixmap(QPixmap::fromImage(imgScaled));
    }else
    {
        qDebug()<<"image_1 is empty";
    }
}
```

图像采集是机器视觉一个非常重要的任务,首先通过建立相机的类和相机的序列号识别不同的相机;然后设定曝光值。需要注意的是,这里不能使用默认的曝光值,因为默认的曝光值是从零开始的,必须设定一个最小的曝光值(如99),否则相机没有办法实现改变曝光值的功能。

6.3 环形分区光源合成算法

利用分区光源在不同地方点亮的方式完成图像的采集。在全亮模式下,图像采集的效果如图6-6所示。

算法的合成分为两个部分,一部分是2D图像的合成效果,如图6-7所示;另一部分是2.5D图像的合成效果,如图6-8所示。接下来首先介绍2D图像的合成。

图 6-6　全亮模式下图像采集的效果

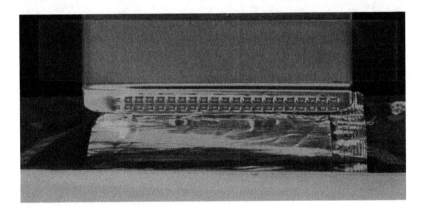

图 6-7　五张图像合成的 2D 图像

图 6-8　五张合成的 2.5D 图像

```
cv :: Mat rawImg = 0.125 * (rawImg2-rawImg1) + 0.125 * (rawImg4-rawImg3) + 0.125 * (rawImg6-
rawImg5) + 0.125 * (rawImg8-rawImg7) + 0.5 * (rawImg10-rawImg9);
cv :: imshow("rawImg", rawImg);
cv :: waitKey(0);
```

从能量的角度来讲,前面四个图像只有四分之一的区域亮,第五张图像是全亮,那么能量分布就是 0.125,0.125,0.125,0.125,0.5。如果按照这个方式来合成的话,我们可以看到,合成图像的能量比我们全亮的图像要暗一些,这是因为前面图像的能量本来就是 0.125 的能量分布,所以需要将这个进行参数调整。

```
double rate_mutily = 0.618;
    cv :: Mat rawImg = rate_mutily * (rawImg2-rawImg1) + rate_mutily * (rawImg4-rawImg3) + rate_
    mutily * (rawImg6-rawImg5) + rate_mutily * (rawImg8-rawImg7) + (1-rate_mutily) * (rawImg10-
    rawImg9);
    cv :: imshow("rawImg", rawImg);
    cv :: waitKey(0);
```

我们可以观察到,图像的变化如图 6-9、图 6-10 和图 6-11 所示。

图 6-9　全亮时候的图像

图 6-10　合成的 2D 图像

总体来看,采用 0.618 是将图像的亮度进行了一个提升(从原来的 0.5 提升至 0.618),因为原来在全亮时,图像的亮度有点偏暗,所以需要将图像的亮度进行提升。

采用分时频闪时需要注意,采用光的角度大约为 15~45 度的低角度的方式采集,这种方式可以有限地解决光的反射和遮挡的问题,同时会提升图像采集和合成的时间。

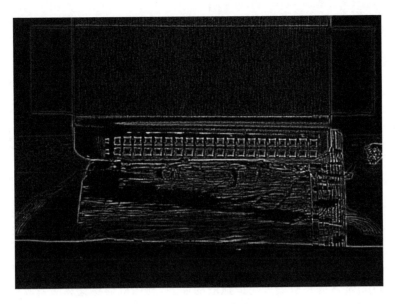

图 6-11　2.5D 合成图像

思 考 题

简答题

1. 在什么条件下，HIK 是可以自己使用 Basler 相机？
2. 分时频闪主要解决哪两个问题？
3. 函数接口、相机接口和相机调用之间的关系是什么？

第 7 章

C++和 OpenCV 视觉九点标定和五点旋转标定

7.1 背景介绍

在机器视觉中,常常需要机器人完成抓取的任务,利用机器视觉算法完成机器视觉的建模和标定是机器人应用中的一个常见的应用。如何建立一套机器人抓取的视觉建模与标定系统,是机器人应用中需要解决的重要问题。基于 C++和 OpenCV 的一套视觉建模系统与标定的方法,能实现机器人快速建模与标定,是解决问题的核心。

建立机器人抓取的视觉建模与标定系统,我们需要给机器人以下数据:①模板位的位置、角度和当前抓取物体的位置、角度;②九点标定后像素点的大小;③旋转标定后,机器人的旋转中心位置。为了获取这些数据,大多数公司采用商业软件来完成(如美国的 VisualPro 和德国的 Halcon),这导致大多数应用公司和自动化装备公司没有自主开发能力。所以,开发一套通用的机器人抓取的视觉建模与标定方式,有很强的迫切性和必要性。

7.2 视觉建模

视觉建模,即选择一个比较好的位置作为机器人抓取的标准位置,将后面物体需要抓取的当前位置与角度计算出来,"告诉"机器人。九点标定,即采用一个标准块,让机器人运行一段固定的距离,本次实验的固定距离是 14 mm。那么,我们首先采集九张图片,如图 7-1 所示。

根据九点标定,机器人每次移动的距离是 14 mm,那么首先需要确定这九点的位置。如图 7-2 所示。

1x	1y	2x	2y	3x	3y
1 661.0	778.0	1 662.0	664.0	1 773.268 799	664.497 437

4x	4y	5x	5y	6x	6y
1 775.0	779.0	1 774.0	892.0	1 657.705 078	690.756 104

7x	7y	8x	8y	9x	9y
1 547.0	691.0	1 548.0	777.0	1 549.0	664

第 7 章 C++和 OpenCV 视觉九点标定和五点旋转标定

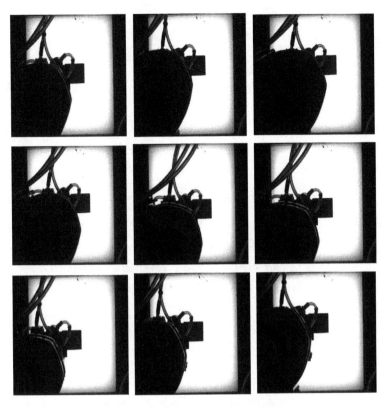

图 7-1 九个位置的分别采集图像

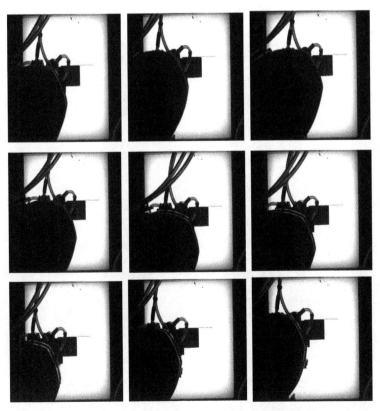

图 7-2 九个点的直角点的位置显示图像

像素标点的结果如下。

778.0－664.0＝114.0

14/114.0＝0.122 8 mm

像素点的大小决定了检测的精度,通常情况下,将像素点大小的 2 倍作为测量的精度。本次测量精度为 0.122 8×2 mm。

机器人的九点坐标的位置与图像的九点位置之间的转换,称为九点标定。首先,对于九点标定而言,我们使用到的是 OpenCV 中的 estimateRigidTransform 函数。

函数定义如下。

```
Mat estimateRigidTransform(InputArraysrc,InputArraydst,boolfullAffine)
```

前两个参数可以是 src＝srcImage(变换之前的图片 Mat),dst＝transImage(变换之后的图片 Mat);也可以是 src＝array(变换之前的关键点 Array),dst＝array(变换之后的关键点 Array)。第三个参数是 1(全仿射变换,包括 rotation, translation, scaling, shearing, reflection)。其主要原理为:如果我们有一个点变换之前是[x,y,1],变换后是[x',y',1]则 fullAffine 表示如下

```
TX = Y
```

标定步骤如下。

1. 准备一块标定板。如果条件不允许,可以在白纸上画九个圆代替。

2. 固定好相机位置和机械手位置,并将标定针固定在机械手上,标定针被固定好后不能再移动。注意,一定要保持标定针与夹手或吸盘内的工具同一位置高度。

3. 将标定板放到相机下方,位置区域要与机械手工作的区域一样,高度也尽量保持一致。这是标定准确度的关键。

4. 调整好相机焦距后拍照,然后识别 9 个圆圆心的坐标并记录。

5. 将机械手依次移动到 9 个圆的圆心位置,记下机械手坐标。

完成以上五步,我们会得到两个点集,一个是 9 个圆的圆心坐标(points_camera),另一个是 9 个圆心对应的机械手坐标(points_robot)。如下所示。

```
Mat warpMat;
vector<Point2f>points_camera;
vector<Point2f>points_robot;
vector<Point2f>points_camera;
vector<Point2f>points_robot;
warpMat = estimateRigidTransform(points_camera, points_robot, true);
A = warpMat.ptr<double>(0)[0];
B = warpMat.ptr<double>(0)[1];
C = warpMat.ptr<double>(0)[2];
D = warpMat.ptr<double>(1)[0];
E = warpMat.ptr<double>(1)[1];
F = warpMat.ptr<double>(1)[2];
```

得出来的 6 个 double 类型的参数就是我们此次标定最终得到的标定参数。

我们代入检测得到的图像坐标(t_px,t_py),就可以得到与之相对应的机械手坐标(t_rx,t

_ry),如下所示。

```
t_rx = (A * t_px) + B * t_py + C;
t_ry = (D * t_px) + E * t_py + F;
```

至此标定结束,我们可以控制相机拍照进行定位,然后转换成机械手坐标。

7.3 旋转标定

旋转标定的作用是根据五点找到一个污染点后,进行丢弃,然后进行第二次定位。首先,我们需要计算五个点的位置,如图 7-3、图 7-4、图 7-5、图 7-6 和图 7-7 所示,五点旋转标点的结果如图 7-8 所示。

图 7-3 旋转图像 1 直角点的位置

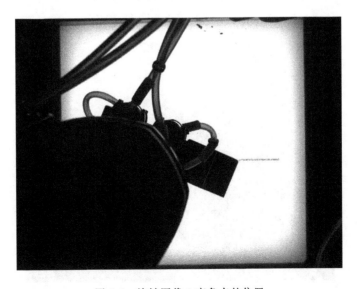

图 7-4 旋转图像 2 直角点的位置

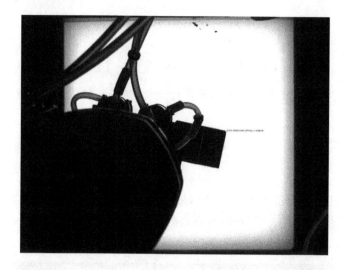

图 7-5 旋转图像 3 直角点的位置

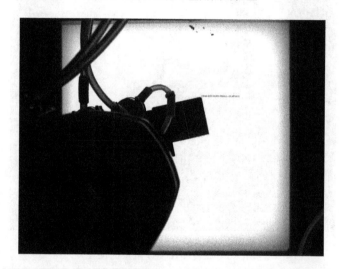

图 7-6 旋转图像 4 直角点的位置

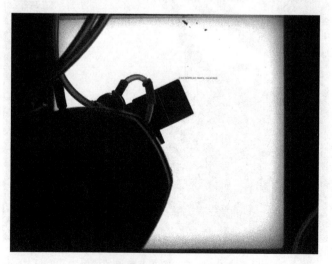

图 7-7 旋转图像 5 直角点的位置

第 7 章 C++和 OpenCV 视觉九点标定和五点旋转标定

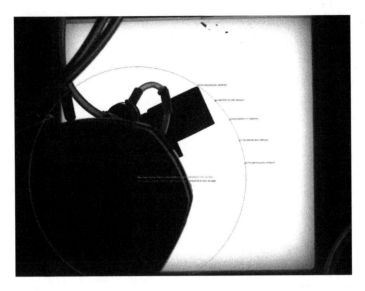

图 7-8 旋转图像所有直角点的位置

九点标定和五点旋转标定是机器人应用中必须完成的任务。有了九点的图像位置和机器人本身的物理坐标位置后,我们就可以通过变换,获得位置坐标到图像位置坐标的转换。确定旋转坐标后,我们可以根据九点转换的原理,获得机器人的物理的旋转中心,这样就可以进行机器人抓取了。

7.4 代码分析

代码的编码环境是在 Win 10 64 位的操作系统下,将 VS 2019 作为编译系统,将开源图像库 OpenCV 4.5.3 作为开发库来完成的。如下所示。

```
/* five points
* 1(1415.322632,521.525879)
* 2(1548.605103,636.386841)
* 3(1658.0,777.0)
* 4(1731.692261,934.538452)
* 5(1772.387573,1106.470993)
*/
std::vector<cv::Point2f> fivepoints;
cv::Point2f temp_point1;
temp_point1.x = 1415.322632;
temp_point1.y = 521.525879;
fivepoints.push_back(temp_point1);   //1 point


cv::Point2f temp_point2;
temp_point2.x = 1548.605103;
temp_point2.y = 636.386841;
```

```cpp
fivepoints.push_back(temp_point2);    //2 point

cv :: Point2f temp_point3;
temp_point3.x = 1658.0;
temp_point3.y = 777.0;
fivepoints.push_back(temp_point3);    //3 point

cv :: Point2f temp_point4;
temp_point4.x = 1731.692261;
temp_point4.y = 934.538452;
fivepoints.push_back(temp_point4);    //4 point

cv :: Point2f temp_point5;
temp_point5.x = 1772.387573;
temp_point5.y = 1106.470993;
fivepoints.push_back(temp_point5);    // 5 point
cv :: Point3f tempcircle;
LeastSquareFittingCircle(fivepoints, tempcircle);
std :: cout <<"centerPoint.x = "<< tempcircle.x <<"centerPoint.y = "<< tempcircle.y << std :: endl;
cv :: Point2f center_one_time;
center_one_time.x = tempcircle.x;
center_one_time.y = tempcircle.y;

//计算圆心到5个点的距离
double max_distance = 0.0;
double d_mean_dis = 0.0;
for(int i = 0; i < fivepoints.size(); i++)
{
    double temp_distance = two_points_distance(center_one_time, fivepoints[i]);
    d_mean_dis + = temp_distance;
}

d_mean_dis = d_mean_dis /5.0;

for(int i = 0; i < fivepoints.size(); i++)
{
    double temp_distance = two_points_distance(center_one_time, fivepoints[i]);
    double abs_dis = abs(temp_distance-d_mean_dis);
    if(abs_dis > max_distance)
    {
        max_distance = abs_dis;
    }
```

```cpp
}

//丢弃最远距离的点。
std::vector<cv::Point2f> fourpointsok;
for(int i = 0; i < fivepoints.size(); i++)
{
    double temp_distance = two_points_distance(center_one_time, fivepoints[i]);
    std::cout << "The temp_distance " << i << "temp_distance = " << temp_distance << std::endl;
    double abs_dis = abs(temp_distance-d_mean_dis);
    if(abs_dis != max_distance)
    {
        fourpointsok.push_back(fivepoints[i]);
        std::cout << "The " << i << "dis = " << abs_dis << std::endl;
    }
}

std::cout << "fourpointsok size = " << fourpointsok.size() << std::endl;

///* cv::Point2f   center_point_better;
//double epsilon_better = 0.01;
//morePointCenter(fourpointsok, center_point_better, epsilon_better);*/

cv::Point3f tempcircle_better;

LeastSquareFittingCircle(fourpointsok, tempcircle_better);
std::cout << "tempcircle_better.x = " << tempcircle_better.x << "tempcircle_better.y = " << tempcircle_better.y << std::endl;

cv::Point2f center_two_time;
center_two_time.x = tempcircle_better.x;
center_two_time.y = tempcircle_better.y;
```

整个方法为：首先计算五个点的圆心，然后比较五个点距圆心的距离，将距离最大的点去掉后，利用剩下的四个点再次计算圆心，就获得机器人在图像坐标下的位置。

思 考 题

简答题

1. 九点标定的数学原理是什么？目的是什么？
2. 五点旋转标定的目的是什么？去掉一个点的原则是什么？

第 8 章

C++和 OpenCV 实现卷绕视觉纠偏

8.1 纠偏需求分析

在锂电行业中,卷绕机是一个非常重要的智能装备,是为了提高卷绕的质量而被引入视觉纠偏系统。在实际的生成过程中,由于视觉部分有可能受到外界干扰(比如激光焊接产生的高强光),这时就会对图像采集后的算法提出更高的要求。算法既要满足图像在低曝光条件下采集的图像(图像会比较暗),又要满足在高曝光条件下采集的图像(图像会比较亮)。由于光照的原因,有可能光照会很强,导致图像出现过曝光的问题。视觉纠偏系统的原图采集和测试结果分别如图 8-1 和图 8-2 所示。

图 8-1 原图采集 1

图 8-2 测试图 1

在电池卷绕的图像采集时,中间黑色部分被称为 AT9,由于受机械空间和图像采集时间的限制,图像效果会发生一些变化,原图和测试图分别如图 8-3 和图 8-4 所示。

从图 8-3 中可以看到,左边的基准线是比较容易找到的,中间的线是 AT9 的特征,由于 AT9 特征不是很明显,需要用增强的算法进行增强后,进行提取特征。

图 8-3　原图采集 2

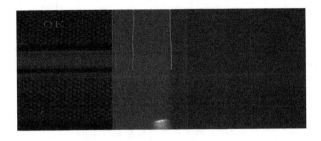

图 8-4　测试图 2

我们接着观察图 8-5 和图 8-6。

图 8-5　原图采集 3

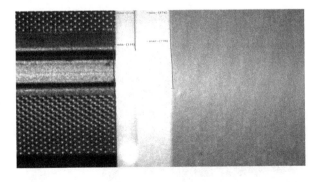

图 8-6　测试图 3

在寻找基准线时,如果按照下面的基准线寻找基准线,上面的基准线就不是很准确,造成这样的原因是采集图像过曝。通常情况下,无论是厂内调试设备,还是厂外调试设备,调整光源亮度和相机的曝光时间两个参数时,尽量避免过曝。

8.2 图像处理流程

从计算机视觉或者图像处理来看,图像处理流程可以分为两个阶段:前处理和后处理。无论图像处理的难易程度如何,大体上来讲,可以用一个比较有共性的流程图来表示。如图8-7所示。

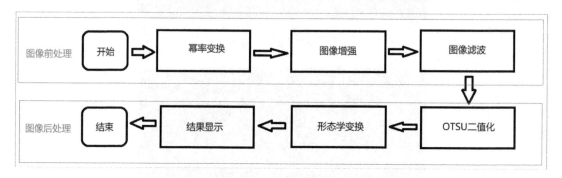

图8-7 图像算法处理流程框架图

图像前处理又可分为三个阶段:幂率变换、图像增强和图像滤波。需要处理的原始图像如图8-8所示,测试结果图如图8-9所示。

图8-8 需要处理的原始图像

图8-9 获得的测试结果图

8.3 左边基准线

图像前处理的代码如下所示。

```
/////////////////////
//幂率变换
//φ>1：处理漂白的图片,进行灰度级压缩
//φ<1：处理过黑的图片,对比度增强,使得细节看的更加清楚
    //构造输出图像
float fGamma = 6.5;
float fC = 1.0;

cv :: Mat matInput;
ROI_A.copyTo(matInput);
cv :: Mat matOutput = cv :: Mat :: zeros(matInput.rows, matInput.cols, matInput.type());
MyGammaCorrection(matInput, matOutput, 0.4);
cv :: imshow("幂率变换", matOutput);
cv :: waitKey(0);
//高斯滤波 降噪
cv :: Mat gaussImg;
cv :: GaussianBlur(matOutput, gaussImg, cv :: Size(3, 3), 0);
cv :: imshow("高斯滤波", gaussImg);
cv :: waitKey(1);
//边缘增强
/*
cv :: Mat kernel = (cv :: Mat_<float>(3, 3)<<1, 1, 1, 1, -8, 1, 1, 1, 1);
cv :: Mat imgLaplance;
cv :: filter2D(gaussImg, imgLaplance, CV_32F, kernel, cv :: Point(-1, -1), 0, cv :: BORDER_DEFAULT);
cv :: imshow("imgLaplance", imgLaplance);
cv :: waitKey(WAITTIEM);
gaussImg.convertTo(gaussImg, CV_32F);
cv :: Mat resultImg = gaussImg-imgLaplance;
resultImg.convertTo(resultImg, CV_8UC1);
cv :: imshow("边缘增强", resultImg);
cv :: waitKey(WAITTIEM);
*/
```

图像后处理的代码如下所示。

```
cv :: Mat binary;
cv :: threshold(gaussImg, binary, 0, 255, cv :: THRESH_OTSU);
cv :: imshow("binary", binary);
cv :: waitKey(1);
```

```cpp
//形态学变换
cv::Mat MM_Img;
cv::Mat kernel_MM = getStructuringElement(cv::MORPH_RECT, cv::Size(1, 1), cv::Point(-1, -1));
cv::morphologyEx(binary, MM_Img, cv::MORPH_CLOSE, kernel_MM);//开操作
/*cv::Mat kernel_MM2 = getStructuringElement(cv::MORPH_RECT, cv::Size(2, 25), cv::Point(-1, -1));
cv::dilate(MM_Img,MM_Img,kernel_MM2);*/
cv::imshow("MM_Img", MM_Img);
cv::waitKey(1);

cv::Mat contours_find;
cv::Canny(MM_Img, contours_find, 155, 350);
cv::imshow("contours_find", contours_find);
cv::waitKey(1);
```

将直线找到并且画出来的代码如下所示。

```cpp
typedef struct
{
    cv::Point start_point;
    cv::Point end_point;

}TWO_POINTS;
std::vector<TWO_POINTS> Line_X_Points;
std::vector<TWO_POINTS> Line_Y_Points;

std::vector<cv::Point> key_Points;

std::vector<cv::Vec2f> lines;
cv::HoughLines(contours_find, lines, 1, PI / 180, 180);
//cv::imshow("cany",contours);
std::vector<cv::Vec2f>::const_iterator it = lines.begin();
while(it != lines.end())
{
    float rho = (*it)[0]; // first element is distance rho
    float theta = (*it)[1]; // second element is angle theta
    if(theta < PI / 4. || theta > 3. * PI / 4.)//~vertical line
    {
        // point of intersection of the line with first row
        cv::Point pt1(rho / cos(theta), 0);
        // point of intersection of the line with last row
```

```cpp
            cv :: Point pt2((rho-color_gray.rows * sin(theta)) / cos(theta), color_gray.rows);
            // draw a white line
            cv :: line(color_gray, pt1, pt2, cv :: Scalar(0,0,255), 1);
            TWO_POINTS temp_two_points;
            temp_two_points.start_point.x = pt1.x;
            temp_two_points.start_point.y = pt1.y;
            temp_two_points.end_point.x = pt2.x;
            temp_two_points.end_point.y = pt2.y;

            Line_Y_Points.push_back(temp_two_points);
            cv :: circle(color_gray, pt1, 3, cv :: Scalar(255, 0, 125), 3, 8);
            key_Points.push_back(pt1);

        }
        else
        { //~horizontal line
            // point of intersection of the
            // line with first column
            cv :: Point pt1(0, rho / sin(theta));
            // point of intersection of the line with last column
            cv :: Point pt2(color_gray.cols,(rho-color_gray.cols * cos(theta)) / sin(theta));
            // draw a white line
            cv :: line(color_gray, pt1, pt2, cv :: Scalar(255,0,0), 1);
            TWO_POINTS temp_two_points;
            temp_two_points.start_point.x = pt1.x;
            temp_two_points.start_point.y = pt1.y;
            temp_two_points.end_point.x = pt2.x;
            temp_two_points.end_point.y = pt2.y;

            Line_X_Points.push_back(temp_two_points);

            cv :: circle(color_gray, pt1, 3, cv :: Scalar(255, 0, 255), 3, 8);

        }
        ++it;
    }
    cv :: imshow("Lines", color_gray);
    cv :: waitKey(1);
```

图像的基准线查找结果如图 8-10 所示。

图 8-10　图像的基准线查找结果

寻找最靠近基准线的左基准线的结果如图 8-11 所示。

图 8-11　图像左基准线结果

8.4　右边基准线

对于右边基准线的处理代码如下所示。

```
cv :: Mat matOutput2 = cv :: Mat :: zeros(matInput.rows, matInput.cols, matInput.type());
MyGammaCorrection(matInput, matOutput2, 0.4);
cv :: imshow("幂率变换2", matOutput2);
cv :: waitKey(1);

//模糊增强
int throDark = mean_value(matOutput2) - 30;
int throMid = mean_value(matOutput2);
int throBright = mean_value(matOutput2) + 30;

cv :: Mat fuzzy_IMG;
```

```cpp
    fuzzy_deal(matOutput2, fuzzy_IMG, throDark, throMid, throBright);
    cv::imshow("模糊增强", fuzzy_IMG);
    cv::waitKey(1);

    //添加底帽变换
    cv::Mat hat_Mat;
    cv::Mat element_set = cv::getStructuringElement(cv::MORPH_RECT, cv::Size(45, 45));
    cv::morphologyEx(fuzzy_IMG, hat_Mat, cv::MORPH_TOPHAT, element_set);
    cv::imshow("hat_Mat", hat_Mat);
    cv::waitKey(1);

    cv::Mat sub_IMG;
    cv::subtract(fuzzy_IMG, hat_Mat, sub_IMG);
    cv::imshow("sub_IMG", sub_IMG);
    cv::waitKey(1);

    //高斯滤波 降噪
    cv::Mat gaussImg2;
    cv::GaussianBlur(sub_IMG, gaussImg2, cv::Size(5, 199), 0);
    cv::imshow("高斯滤波 2", gaussImg2);
    cv::waitKey(1);
```

右边基准线的后处理代码如下所示。

```cpp
    //SOBLE
    cv::Mat grad_x, grad_y;
    cv::Sobel(gaussImg2, grad_x, CV_16S, 1, 0, 3, 1, 0, cv::BORDER_DEFAULT);
    cv::Sobel(gaussImg2, grad_y, CV_16S, 0, 1, 3, 1, 0, cv::BORDER_DEFAULT);
    cv::convertScaleAbs(grad_x, grad_x);
    cv::convertScaleAbs(grad_y, grad_y);
    cv::imshow("grad_x", grad_x);
    cv::waitKey(1);
    cv::imshow("grad_y", grad_y);
    cv::waitKey(1);

    cv::Mat binary_R;
    cv::threshold(grad_x, binary_R, 0, 255, cv::THRESH_OTSU);
    cv::imshow("binary_R", binary_R);
    cv::waitKey(1);
```

显示的方法代码如下所示。

```cpp
//将过滤掉后,只是剩下最大的区域
    //start
    std::vector<std::vector<cv::Point>> g_vContoursR;


    std::vector<cv::Vec4i> g_vHierarchyR;
```

```cpp
cv::findContours(binary_R, g_vContoursR, g_vHierarchyR, CV_RETR_EXTERNAL, cv::CHAIN_APPROX_SIMPLE);
std::sort(g_vContoursR.begin(), g_vContoursR.end(), ContoursSizeSortFun);

//按照从大到小排序
for(int i = 0; i < g_vContoursR.size(); i++) {
    std::cout << "i = " << i << "size = " << cv::contourArea(g_vContoursR[i]) << std::endl;
}


std::vector<std::vector<cv::Point>> LR_ContoursR;
for(int i = 0; i < 1; i++) {
    LR_ContoursR.push_back(g_vContoursR[i]);
}
//左右轮廓按照X轴进行排序
//std::sort(LR_Contours.begin(), LR_Contours.end(), ContoursSortXFun);

//L to R
//按照从大到小排序
for(int i = 0; i < LR_ContoursR.size(); i++) {
    std::cout << "i = " << i << "LR_Contours size = " << cv::contourArea(LR_ContoursR[i]) << std::endl;
}

//2020.04.30 显示

//将左右两个都显示出来
for(unsigned i = 0; i < LR_ContoursR.size(); i++)
{
    cv::RotatedRect box = minAreaRect(LR_ContoursR[i]);
    cv::Point2f boxPoints[4];
    box.points(boxPoints);
    cv::Point2f pointA = midpoint(boxPoints[0], boxPoints[1]);
    cv::Point2f pointB = midpoint(boxPoints[1], boxPoints[2]);
    cv::Point2f pointC = midpoint(boxPoints[2], boxPoints[3]);
    cv::Point2f pointD = midpoint(boxPoints[3], boxPoints[0]);
    cv::circle(color_gray, pointA, 2, cv::Scalar(0, 0, 255));
    cv::circle(color_gray, pointB, 2, cv::Scalar(0, 0, 255));
    cv::circle(color_gray, pointC, 2, cv::Scalar(0, 0, 255));
    cv::circle(color_gray, pointD, 2, cv::Scalar(0, 0, 255));
    cv::line(color_gray, pointA, pointC, cv::Scalar(255, 0, 0));
    cv::line(color_gray, pointD, pointB, cv::Scalar(255, 0, 0));
    double dWidth = getDistance(pointA, pointC);
```

```
    double dHeight = getDistance(pointD, pointB);
    cv :: putText(color_gray, cv :: format("( %2d, %2d)", i, i), boxPoints[2], cv :: FONT_
    HERSHEY_COMPLEX, 0.5, cv :: Scalar(0, 0, 255));

    for(int i = 0; i <= 3; i++)
    {
        cv :: line(color_gray, boxPoints[i], boxPoints[(i+1) % 4], cv :: Scalar(0, 255,
        0));
    }

}
cv :: imshow("color_gray", color_gray);
cv :: waitKey(1);
```

这样就获得了右边基准线的结果,如图 8-12 所示。

图 8-12 右边基准线的效果图

8.5 中间特征线

AT9 中间特征的寻找是采用图像金字塔的算法来提取的,具体的代码如下所示。

```
//1
    cv :: Mat imagemen1;
    cv :: blur(rawImg, imagemen1, cv :: Size(31, 31));
    cv :: imshow("imagemen1", imagemen1);
    cv :: waitKey(1);

    //
    cv :: Mat diff1;
    cv :: subtract(imagemen1, rawImg, diff1);
    cv :: imshow("diff1", diff1);
```

```cpp
    cv::waitKey(1);

    //2
    cv::Mat imagemen2;
    cv::Mat down_img2;
    //降采样(zoom out 缩小)
    pyrDown(rawImg, down_img2, cv::Size(rawImg.cols / 2, rawImg.rows / 2));

    imshow("zoomdonw2", down_img2);
    cv::blur(down_img2, imagemen2, cv::Size(15, 15));
    cv::imshow("imagemen2", imagemen2);
    cv::waitKey(1);

    //
    cv::Mat diff2;
    cv::subtract(imagemen2, down_img2, diff2);
    cv::imshow("diff2", diff2);
    cv::waitKey(1);

    //上采样(zoom in 放大)
    cv::Mat diff2UP;
    pyrUp(diff2, diff2UP, cv::Size(diff2.cols * 2, diff2.rows * 2));
    imshow("zoomUP2", diff2UP);
    cv::waitKey(1);
    //3
    cv::Mat imagemen3;
    cv::Mat down_img3;
    //降采样(zoom out 缩小)
    pyrDown(down_img2, down_img3, cv::Size(down_img2.cols / 2, down_img2.rows / 2));
    imshow("zoomdonw3", down_img3);
    cv::blur(down_img3, imagemen3, cv::Size(7, 7));
    cv::imshow("imagemen3", imagemen3);
    cv::waitKey(1);

    //
    cv::Mat diff3;
    cv::subtract(imagemen3, down_img3, diff3);
    cv::imshow("diff3", diff3);
    cv::waitKey(1);

    //上采样(zoom in 放大)
    cv::Mat diff3UP;
    pyrUp(diff3, diff3UP, cv::Size(diff3.cols * 2, diff3.rows * 2));
```

```cpp
pyrUp(diff3UP, diff3UP, cv::Size(diff3UP.cols * 2, diff3UP.rows * 2));
imshow("zoomUP3", diff3UP);
cv::waitKey(1);
//4
cv::Mat imagemen4;
cv::Mat down_img4;
//降采样(zoom out 缩小)
pyrDown(down_img3, down_img4, cv::Size(down_img3.cols / 2, down_img3.rows / 2));

imshow("zoomdonw4", down_img4);
cv::blur(down_img4, imagemen4, cv::Size(3, 3));
cv::imshow("imagemen4", imagemen4);
cv::waitKey(1);

//
cv::Mat diff4;
cv::subtract(imagemen4, down_img4, diff4);
cv::imshow("diff4", diff4);
cv::waitKey(1);

//上采样(zoom in 放大)
cv::Mat diff4UP;
pyrUp(diff4, diff4UP, cv::Size(diff4.cols * 2, diff4.rows * 2));
pyrUp(diff4UP, diff4UP, cv::Size(diff4UP.cols * 2, diff4UP.rows * 2));
pyrUp(diff4UP, diff4UP, cv::Size(diff4UP.cols * 2, diff4UP.rows * 2));
imshow("zoomUP4", diff4UP);
cv::waitKey(1);

//diff4UP SIZE CHANGE
cv::resize(diff4UP,diff4UP,cv::Size(diff3UP.cols,diff3UP.rows));

cv::Mat diff;
cv::add(diff1 + diff2UP + diff3UP, diff4UP, diff);

cv::imshow("diffOK", diff);
cv::waitKey(1);
```

整个中间特征的提取采用了四层图像金字塔的算法,具体步骤如下。

第一步:基于左定位和右定位,获得AT9的感兴趣区域(ROI),如图8-13所示。

第二步:利用四层图像金字塔算法提取特征后,进行二值化。

第三步:进行形态学处理。

第四步:结果显示。

图 8-13 中间图像 ROI 获取图像效果

采用图像的形态学,即后处理方法的代码如下。

```
////MM OPERATOR
////形态学变换
//
cv :: Mat kernel_L = getStructuringElement(cv :: MORPH_RECT, cv :: Size(15, 15), cv :: Point(-1,-1));
cv :: morphologyEx(dst_L, dst_L, cv :: MORPH_CLOSE, kernel_L);//开操作
cv :: imshow("MM_LOK", dst_L);
cv :: waitKey(1);
    //MM 细化
cv :: Mat thindst;
thinIMG(dst_L, thindst);;
cv :: imshow("细化操作", thindst);
cv :: waitKey(1);


//show_result
std :: vector< cv :: Point > result_points;
//
int height = thindst.rows; //binary_F.rows
```

```cpp
int width = thindst.cols;


for(int row = 0; row < height; row++)
{
for(int col = 0; col < width; col++)
    {
    if(thindst.at<uchar>(row, col) == 255)
        {
            cv::Point tempPoint;
            tempPoint.x = col + Max_X + 50;
            tempPoint.y = row;
            result_points.push_back(tempPoint);
        }

    }
}

for(int i = 0; i < result_points.size(); i++)
{
    cv::circle(show_result, result_points[i], 1, cv::Scalar(255, 0, 255), 1, 8);
}

cv::imshow("show_result", show_result);
cv::waitKey(1);
```

通过本次图像算法的学习,我们知道针对不同的图像采集需求,需要采用不同的算法。作为智能装备的图像算法,过检率、漏检率和处理时间是衡量一个算法好坏的最重要的三大指标。

思 考 题

简答题
1. 图像处理的基本流程是什么？
2. 衡量一个图像算法好坏的三大指标是什么？

第 9 章

C++和 OpenCV 实现键盘缺陷检测

9.1 需求分析

本次检测是针对笔记本计算机的生成过程中，键盘的缺陷检测展开的。键盘缺陷的种类包括：混键（如 Q 键和 O 键位置放置错误）、缺键（如 A 键脱落）、语言错误（如日文写成了韩文）、印刷错误和印刷不良等。

9.2 图像 ROI

采集键盘的原始图像。采用 2000 万像素彩色相机和四个条形光源来采集笔记本计算机的图像，如图 9-1 所示。总体目标是，需要根据笔记本计算机的型号对应的键盘建模。这样，在生产过程中，我们可以根据产品的工单、对应的建模进行切换。

图 9-1 采集键盘的原始图像

首先进行特征点定位,获取 ROI 的图像,如图 9-2 所示。

图 9-2　图像的特征点定位

这样就可以获得键盘的 ROI 图像,如图 9-3 所示。

图 9-3　获得键盘的 ROI 图像

9.3　分区建模

将整个键盘分为三个区域,第一个是主键盘区域,第二个是数字键盘区域,第三个是功能键盘区域。将这三个区域分割完毕后,分别定位提取对应的区域并存储到 XML 文件中。

通过图像的分割,将整个键盘分为了主键盘区域(如图 9-4 所示),数字键盘区域(如图 9-5 所示),功能键盘区域(如图 9-6 所示)。

图 9-4 主键盘图像

图 9-5 数字键盘图像

图 9-6 功能键盘图像

9.4 模板 XML

为了存储每个键的位置,将每个键的上下左右和中心点共计五个点进行存储,同时也对每个键的角度进行存储。建立结构体代码如下所示。

```
struct fivePoints
{
cv :: Point2f centerPoint;
cv :: Point2f leftupPoint;
cv :: Point2f rightupPoint;
cv :: Point2f rightdnPoint;
cv :: Point2f leftdnPoint;
double angle_get;
};
```

将 ROI 第一区域的第一行的键盘进行处理的代码如下所示。

```cpp
std::vector<fivePoints> roi_11;
for(int i = 0; i < ImgKeyPoints.size(); i++)
{
    if(abs(ImgKeyPoints[i].centerPoint.y - 37) < 15)
    {
        roi_11.push_back(ImgKeyPoints[i]);
    }
}
std::sort(roi_11.begin(), roi_11.end(), ROISortXFun);
```

将每个键盘的键的数据进行存放的代码如下所示。

```cpp
//PUT AHEAD
    for(int i = 0; i < roi_11.size(); i++)
    {
        cv::Mat temp_ROI = BIGROI1(cv::Rect(roi_11[i].leftupPoint.x + 2, roi_11[i].leftupPoint.y + 2, roi_11[i].rightupPoint.x-roi_11[i].leftupPoint.x-7, roi_11[i].leftdnPoint.y-roi_11[i].leftupPoint.y-9));
        cv::imshow("temp_ROI", temp_ROI);
        cv::waitKey(0);
        Mat_ROI11.push_back(temp_ROI);
        BASE_DATA temp_base_data;
        temp_base_data.x_position = roi_11[i].centerPoint.x;
        temp_base_data.y_postion = roi_11[i].centerPoint.y;
        temp_base_data.angle_computer = roi_11[i].angle_get;
        Data_ROI11.push_back(temp_base_data);//Data_ROI11;
    }
```

新建一个 XML 文件代码如下所示。

```cpp
cv::FileStorage fs("keyboard_11.xml", cv::FileStorage::WRITE);//创建 XML 文件
    if(!fs.isOpened())
    {
        std::cerr<<"failed to open "<<"keyboard_11.xml"<< std::endl;
    }
```

将键盘的数据进行存储的代码如下所示。

```cpp
//每个区域有多少个小的图片或字符键
        std::map<std::string, int> ROI_data;
        //ROI_data.insert(std::map<std::string,int>::value_type("ROI4", Mat_ROI4.size()));

        ROI_data.insert(std::map<std::string, int>::value_type("ROI11", Mat_ROI11.size()));

        fs<<"MAP"<<"{";//注意要用到大括号
        for(std::map<std::string,int>::iterator it = ROI_data.begin(); it != ROI_data.end(); it++)
```

```cpp
        {
            fs << it->first << it->second;
        }
    fs << "}";

//fs << "ROI11" << "["; //注意要有中括号
//Map 数据读取:std::vector<BASE_DATA> Data_ROI11;//存储第1区域的x,y,angle,
for(int i = 0; i < Data_ROI11.size(); i++)
{
    std::map<std::string, double> data;
    data.insert(std::map<std::string, double>::value_type("x_postion", Data_ROI11[i].x_position));
    data.insert(std::map<std::string, double>::value_type("y_postion", Data_ROI11[i].y_postion));
    data.insert(std::map<std::string, double>::value_type("angle_correct", Data_ROI11[i].angle_computer));
    fs << "Data_ROI11" + std::to_string(i) << "{";//注意要用到大括号
    for(std::map<std::string, double>::iterator it = data.begin(); it != data.end(); it++)
    {
        fs << it->first << it->second;
    }
    fs << "}";
}
for(int i = 0; i < Mat_ROI11.size(); i++)
{
    fs << "Mat_ROI11" + std::to_string(i); // 注意要有中括号
    fs << Mat_ROI11[i];
}
fs.release();
```

9.5 缺陷算法

键盘的缺陷算法采用SSIM算法来完成,代码如下所示。

```cpp
int SSIM_CPU(cv::Mat src1, cv::Mat src2, double& like_degree)
{
    if(src1.empty() || src2.empty())
    {
        printf("can ont load images....\n");
        return -1;
    }
    if(src1.channels() != 1)
```

```cpp
{
    cv :: cvtColor(src1, src1, CV_BGR2GRAY);
}

if(src2.channels() != 1)
{
    cv :: cvtColor(src2, src2, CV_BGR2GRAY);
}

cv :: resize(src1, src1, cv :: Size(src1.size().width / 4, src1.size().height / 4));
cv :: resize(src2, src2, cv :: Size(src2.size().width / 4, src2.size().height / 4));

imshow("image1", src1);
imshow("image2", src2);
cv :: waitKey(0);

const double C1 = 6.5025, C2 = 58.5225;
/*************************** INITS *********************************/
int d = CV_32F;

cv :: Mat I1, I2;
src1.convertTo(I1, d);           // cannot calculate on one byte large values
src2.convertTo(I2, d);

cv :: Mat I2_2 = I2.mul(I2);     // I2^2
cv :: Mat I1_2 = I1.mul(I1);     // I1^2
cv :: Mat I1_I2 = I1.mul(I2);    // I1 * I2

/*********************** END INITS **********************************/
cv :: Mat mu1, mu2;   // PRELIMINARY COMPUTING
GaussianBlur(I1, mu1, cv :: Size(3, 3), 0.5);
GaussianBlur(I2, mu2, cv :: Size(3, 3), 0.5);

cv :: Mat mu1_2 = mu1.mul(mu1);
cv :: Mat mu2_2 = mu2.mul(mu2);
cv :: Mat mu1_mu2 = mu1.mul(mu2);

cv :: Mat sigma1_2, sigma2_2, sigma12;

GaussianBlur(I1_2, sigma1_2, cv :: Size(3, 3), 0.5);
sigma1_2 -= mu1_2;

GaussianBlur(I2_2, sigma2_2, cv :: Size(3, 3), 0.5);
```

```cpp
    sigma2_2 -= mu2_2;

    GaussianBlur(I1_I2, sigma12, cv::Size(3, 3), 0.5);
    sigma12 -= mu1_mu2;

    /////////////////////////////////////FORMULA ////////////////////////////////////////
    cv::Mat t1, t2, t3;

    t1 = 2 * mu1_mu2 + C1;
    t2 = 2 * sigma12 + C2;
    t3 = t1.mul(t2);                // t3 = ((2 * mu1_mu2 + C1).*(2 * sigma12 + C2))

    t1 = mu1_2 + mu2_2 + C1;
    t2 = sigma1_2 + sigma2_2 + C2;
    t1 = t1.mul(t2);                // t1 = ((mu1_2 + mu2_2 + C1).*(sigma1_2 + sigma2_2 + C2))

    cv::Mat ssim_map;
    divide(t3, t1, ssim_map);       // ssim_map = t3./t1;

    cv::Scalar mssim = mean(ssim_map);  // mssim = average of ssim map

    std::cout<<"mssim = "<<mssim<<std::endl;

    like_degree = mssim[0];
    return 0;
}
```

基于 SSIM 算法的缺陷检测，根据客户提供的样本缺陷及检测的效果来看，在将相似度阈值设置为 0.9 的情况下，所有的缺陷样本都能检测出来。

思 考 题

简答题

1. 图像检测的精度是指的什么？
2. SSIM 匹配算法的原理是什么？

第 10 章
SIFT 和 ANN 实现任意顺序图像的合成

10.1 算法原理

如果一个样品的长度太长,一次拍照很难完成,所以需要分别拍三次、五次或七次才能完成,那么我们需要采用多张合成的方法完成图像的合成。图像匹配方法的数学原理是两个图像之间的欧氏距离,公式如下:

$$d(R_i, S_i) = \sqrt{\sum_{j=1}^{128}(r_{ij}-s_{ij})^2} \tag{1}$$

$$d(R_i, S_i) = \sum_{j=1}^{128}|r_{ij}-s_{ij}| \tag{2}$$

在不同配置参数的条件下的效果不一样,如图 10-1 所示。为了更深入地分析问题,在此引入 k-d 树。在计算机科学中,k-d 树(k-dimensional 树的缩写)是一种空间分区数据结构,用于在 k-dimensional 空间中组织点。三维 k-d 树如图 10-2 所示。

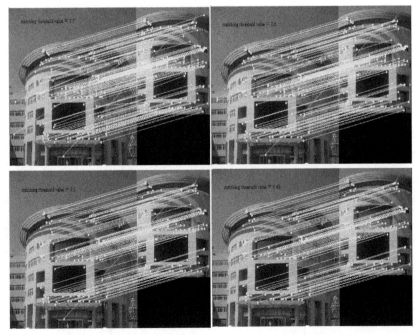

图 10-1 基于匹配参数 0.7,0.6,0.5,0.45 等不同阈值的效果

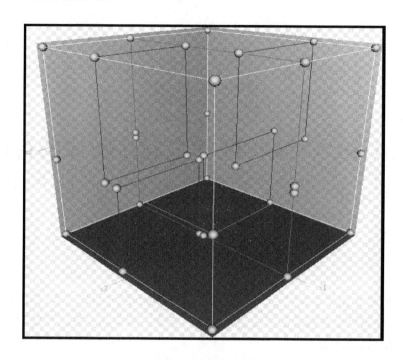

图 10-2 三维 k-d 树

事实上,k-d 树数据结构是基于将空间递归细分为不相交的区域,称为细胞。为了重建二维搜索点,图 10-3 显示了桶大小为 1 的 k-d 树和相应的空间分解,k-d 树的拆分规则完全不同。

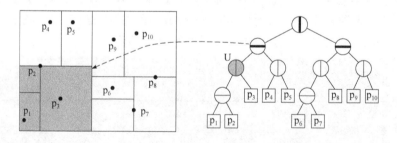

图 10-3 桶大小为 1 的 k-d 树及其相应的空间分解

第一种称为标准 k-d 树拆分规则,拆分维度是最大排列的维度。第二种是一个简单的分割规则,它保证单元格具有有界的纵横比,称为中点分割规则。第三种是对中点分割规则的简单修改,该规则在典型数据集上表现最好。所以,我们应该选择第三种规则的默认规则。

与 k-d 树相比,人工神经网络(Artificial Neural Network,ANN)支持 k-d 树数据结构,为高度聚集的数据集提供更强的健壮性。分裂方法基本上与 k-d 树相同,但 k-d 树的收缩规则仅适用于 k-d 树。如果点低于桶大小,则会为点之间的关系创建叶节点。否则,我们将调用决定是拆分还是收缩的过程。最后,对点进行细分,并在两个子集上递归调用该过程以形成子对象。

10.2 特征寻找

图像根据尺度不变特征转换(Scale-Invariant Feature Transform,SIFT)寻找特征后,进行图像合成的代码如下所示。

```c
#include "stitch.h"
#include "io.h"
#include "time.h"
#include <stdio.h>
#include <stdlib.h>

#define MaxPics 45
#define MaxOut 3

#pragma  comment(lib,"ANN.lib")

int Color[3][8] = {20, 20, 100, 100, 150,150, 230,230,100,150, 150, 230, 20, 230, 100, 20,20,
                   150, 230, 20,150, 230, 100, 100};

double FOCAL = 680;
double HH = 2;
int IntialImageID = 0;
double LOWRATIO = 0.0;//0.07; //0.095;
int MODE = 0;
int Manuel = 0;
double ManuelShift = 0.1259;//0.125920;
int filenum;
void buildKDTree(IplImage * inImg, vector<Feature>* keypoints,ANNkd_tree * *kdTree);
void findNextImage(IplImage * inImg[MaxPics],ANNkd_tree * kdTree[MaxPics],vector<Feature>
keypoints[MaxPics],int nowID, int &NextID, int Flag[MaxPics],CvMat * bestHomo);
void seqNextImage (IplImage * inImg[MaxPics], ANNkd_tree * kdTree[MaxPics], vector<Feature>
keypoints[MaxPics],int nowID, int &NextID,CvMat * bestHomo);
void findTargetHomo(IplImage * inImg[MaxPics],ANNkd_tree * kdTree[MaxPics],vector<Feature>
keypoints[MaxPics],int nowID, int NextID,CvMat * bestHomo);
void seqNextImage2(IplImage * inImg[MaxPics],ANNkd_tree * kdTree[MaxPics],vector<Feature>
keypoints[MaxPics],int nowID, int &NextID,int ind[MaxPics],CvMat * bestHomo);
int readDireadDir(char * path,char fnames[MaxPics][100]);

int readDir(char * path,char fnames[MaxPics][100])
{
    struct _finddata_t c_file;
    int hFile;
```

```cpp
        filenum = 0;

        char fullpath[100];

        sprintf(fullpath,"%s\\*.jpg",path);
        printf(fullpath);
        // Find first .c file in current directory
        if((hFile = _findfirst(fullpath, &c_file)) == -1L)
            printf("No *.jpg files in current directory %s!\n",fullpath);
        else
        {
            printf("Listing of .jpg files in %s directory\n\n",path);
            printf("RDO HID SYS ARC FILE SIZE\n",'');
            printf("---------------------\n",'');
            do {
//              char buffer[30];
                printf((c_file.attrib & _A_RDONLY) ? " Y  " : " N  ");
                printf((c_file.attrib & _A_SYSTEM) ? " Y  " : " N  ");
                printf((c_file.attrib & _A_HIDDEN) ? " Y  " : " N  ");
                printf((c_file.attrib & _A_ARCH)   ? " Y  " : " N  ");
                //ctime_s(buffer, _countof(buffer), &c_file.time_write);
                printf(" %-15s %9ld\n",c_file.name, c_file.size);
                sprintf(fnames[filenum],"%s\\%s",path,c_file.name);
                filenum ++;
            } while(_findnext(hFile, &c_file) == 0);
            _findclose(hFile);
        }
        return filenum;
}

int main(int argc, char * * argv){
    argv[0] = "stitch.exe";
    argv[1] = "test114";
    IplImage * inImg[MaxPics];
    IplImage * inImgc[MaxPics];
    IplImage * inImgFix[MaxPics];//fix 800*600
    IplImage * inImgcFix[MaxPics];//fix 800*600

    vector<Feature> keypoints[MaxPics];
    ANNkd_tree * kdTree[MaxPics];
    //////readFile

    char fdir[MaxPics][100];
```

```
filenum = readDir(argv[1],fdir);
MODE = 0;

for(int i = 0 ; i < filenum ; i++){
    inImg[i] = cvLoadImage(fdir[i],0);
    inImgc[i] = cvLoadImage(fdir[i],1);

        const char * pstrWindowsSrcTitle;
        const char * pstrWindowsDstTitle;
        const char * pstrWindowsSrcTitlec;
        const char * pstrWindowsDstTitlec;

        CvSize czSize;
        CvSize czSizec;
        IplImage * pSrcImage = inImg[i];
        IplImage * pDstImage = NULL;

        IplImage * pSrcImagec = inImgc[i];
        IplImage * pDstImagec = NULL;

        czSize.width = 800;
        czSize.height = 600;

        czSizec.width = 800;
        czSizec.height = 600;

            pDstImage = cvCreateImage(czSize, pSrcImage -> depth, pSrcImage ->
            nChannels);cvResize(pSrcImage, pDstImage, CV_INTER_AREA);

            pDstImagec = cvCreateImage(czSizec, pSrcImagec -> depth, pSrcImagec ->
            nChannels);cvResize(pSrcImagec, pDstImagec, CV_INTER_AREA);
        /*
        cvNamedWindow(pstrWindowsSrcTitle, CV_WINDOW_AUTOSIZE);
        cvNamedWindow(pstrWindowsDstTitle, CV_WINDOW_AUTOSIZE);

        cvNamedWindow(pstrWindowsSrcTitlec, CV_WINDOW_AUTOSIZE);
        cvNamedWindow(pstrWindowsDstTitlec, CV_WINDOW_AUTOSIZE);

        cvShowImage(pstrWindowsSrcTitle, pSrcImage);
        cvShowImage(pstrWindowsDstTitle, pDstImage);

        cvShowImage(pstrWindowsSrcTitlec, pSrcImagec);
        cvShowImage(pstrWindowsDstTitlec, pDstImagec);
```

```
                */
            //cvWaitKey();

            inImgFix[i] = pDstImage;
            inImgcFix[i] = pDstImagec;
}
////////
for(i = 0 ; i< filenum ;i++)
{
    buildKDTree(inImgFix[i],&keypoints[i],&kdTree[i]);
}
IplImage * result[MaxOut], * mark[MaxOut];

for(i = 0 ; i< MaxOut ;i++)
{
    result[i] = cvCreateImage(cvSize(1500,600),IPL_DEPTH_8U,3); //(1500,600)
    mark[i] = cvCreateImage(cvGetSize(result[i]),IPL_DEPTH_8U,1);
    cvSetZero(mark[i]);
}

CvMat * homo;
homo = cvCreateMat(3,3,CV_32FC1);
cvSetIdentity(homo);

CvMat * adjust;
adjust = cvCreateMat(3,3,CV_32FC1);
cvSetIdentity(adjust);

int next = IntialImageID;        /////////////////////////////////stop point
int next2;
int Flag[MaxPics] = {0};
int ind[MaxPics];

CvMat * lastHomo;
lastHomo = cvCreateMat(3,3,CV_32FC1);


int empty = 0;
int outcount = 0;
int flag1 = 0;
double totaltheta;
double totalH;
```

```cpp
if(MODE == 0)
{
    cvSetIdentity(lastHomo);
    printf(" % f\n",cvmGet(lastHomo,1,2));

    while(next != -1)
    {

        printf(" % d  ",next);
        //cvNamedWindow("combine",1);
          //writeCylinderImage5 ( inImgc [ next ], result [ outcount ], 2 * PI, HH, FOCAL, mark
             [outcount],lastHomo);
        Flag[next] = 1;
        //cvShowImage("combine",result[0]);
        //cvWaitKey(0);
        findNextImage(inImg,kdTree,keypoints,next,next2,Flag,homo);
        ind[next] = next2;
        if(next2 == -1)
        {
            findTargetHomo(inImg,kdTree,keypoints,next,IntialImageID,homo);
        }
        next = next2;
        //seqNextImage(inImg,kdTree,keypoints,next,next,homo);

        simpleMul(lastHomo,homo,lastHomo);
        printf(" % f\n",cvmGet(lastHomo,1,2));

    }
    printf("total shift = % f\n",cvmGet(lastHomo,1,2));

    //printf("\nlastHomo = \n");
    totaltheta = cvmGet(lastHomo,0,2);
    totalH = cvmGet(lastHomo,1,2);
    if(Manuel == 1)
        totalH = ManuelShift ;
    totaltheta = totaltheta - 2 * PI;
    totaltheta / = (MaxPics - 1);
    totalH / = (MaxPics - 1);
    //totaltheta * = -1;
    //totalH * = -1;
```

```cpp
        cvmSet(adjust,0,2,totaltheta);
        cvmSet(adjust,1,2,totalH);

        printf("\n");
        cvSetIdentity(lastHomo);
        //printf("%f\n",cvmGet(lastHomo,1,2));
        //simpleMul(lastHomo,adjust,lastHomo);
        next = IntialImageID ;
        for(int i = 0; i < filenum ; i++ )
            Flag[i] = 0;

        while(next != -1)
        {
            printf("%d  ",next);
            //cvNamedWindow("combine",1);


            writeCylinderImage5 ( inImgc [ next ], result [ outcount ], 2 * PI, HH, FOCAL, mark
            [outcount],lastHomo);

            //cvShowImage("combine",result[0]);
            //cvWaitKey(0);

            Flag[next] = 1;

            //findNextImage(inImg,kdTree,keypoints,next,next2,Flag,homo);
            //cvmSet(homo,1,2,0);
            //next = next2;
            seqNextImage2(inImg,kdTree,keypoints,next,next,ind,homo);
            if(next != -1)
            {
                simpleMul(lastHomo,homo,lastHomo);
                //printf("ori = %f\n",cvmGet(lastHomo,1,2));
                //simpleMul(lastHomo,adjust,lastHomo);
            }
            //printf("%f\n",cvmGet(lastHomo,1,2));
        }
        outcount++ ;

    }

    for(i = 0 ; i < outcount ; i++){
        char mixname[100],mixpath[100];
```

```
        sprintf(mixname,"combine_%d",i);
        cvNamedWindow(mixname,1);
        cvShowImage(mixname,result[i]);
        sprintf(mixpath,"panorama_%d.jpg",i+1);
        cvSaveImage(mixpath,result[i]);
    }



    cvWaitKey(0); // very important, contains event processing loop inside


    return 0;
}



void buildKDTree(IplImage * inImg, vector<Feature> * keypoints,ANNkd_tree * * kdTree)
{
    ISIFT * sift;
    sift = new ISIFT();
    //ANNkd_tree * kdTree;

    sift->init();

    //printf("build pyramid..\n");
    sift->buildPyramid(inImg, keypoints);
    sift->release();


    ANNpointArray dataPts;

    int size = keypoints->size();
    dataPts = annAllocPts(size, 128);
    for(int i = 0;i< size;i++ )
    {
        for( int j = 0;j<128;j++ )
        {
            dataPts[i][j] = ( * keypoints)[i].vec[j];
        }
    }
```

```cpp
        * kdTree = new ANNkd_tree(// build search structure
                dataPts, // the data points
                size, // number of points
                128);
    //return kdTree;

}

void findNextImage(IplImage * inImg[MaxPics],ANNkd_tree * kdTree[MaxPics],vector < Feature >
keypoints[MaxPics],int nowID,int &NextID, int Flag[MaxPics],CvMat * bestHomo)
{
    vector < Sample > matchPoints;
    vector < SampleC > matchPointC;

    int match[30000];
    double dist[30000];
    CvMat * homo;
    homo = cvCreateMat(3,3,CV_32FC1);
    int bestCount;
    int bestNext;
    int count;
    bestCount = 0;
    bestNext = -1;
    double bestratio = 0;
    double ratio;


    for(int i = 0; i < filenum ; i ++ )
    {

        if(Flag[i] == 0)
        {
            featureMatch(keypoints[nowID],keypoints[i],kdTree[nowID],match,dist);
            int size2;
            size2 = keypoints[i].size();
            pickMatchPoint(keypoints[nowID],keypoints[i],match,dist,size2,matchPoints);
            CvPoint center;
            CvSize imgSize;
            imgSize = cvGetSize(inImg[nowID]);
            center.x = imgSize.height/2;
            center.y = imgSize.width/2;
            batchToCylinder(matchPoints,center,FOCAL,matchPointC);
```

```cpp
                if(matchPointC.size()>0)
                {
                    Myransac2(matchPointC,homo,1,count);
                    ratio = (double)count /(double)matchPointC.size();
                    //ratio = (double)count / size2;

                    if(ratio > bestratio)
                    //if(count > bestCount)
                    {
                        bestCount = count;
                        bestratio = ratio;
                        bestNext = i;
                        cvmSet(bestHomo,0,2,cvmGet(homo,0,2));
                        cvmSet(bestHomo,1,2,cvmGet(homo,1,2));
                    }
                }
            }
        matchPoints.clear();
        matchPointC.clear();

    }
    NextID = bestNext;
    if(bestratio < LOWRATIO)
        NextID = -1;


    //Flag[bestNext] = 1;

}

void seqNextImage(IplImage * inImg[MaxPics],ANNkd_tree * kdTree[MaxPics],vector < Feature >
keypoints[MaxPics],int nowID,int &NextID,CvMat * bestHomo)
{
    vector < Sample > matchPoints;
    vector < SampleC > matchPointC;
    int match[1000];
    double dist[1000];
    CvMat * homo;
    homo = cvCreateMat(3,3,CV_32FC1);
    int count;

    if((nowID + 1) == filenum)
    {
        NextID = -1;
    }
```

```cpp
        else
        {
            featureMatch(keypoints[nowID],keypoints[nowID + 1],kdTree[nowID],match,dist);
            int size2;
            size2 = keypoints[nowID + 1].size();
            pickMatchPoint(keypoints[nowID],keypoints[nowID + 1],match,dist,size2,matchPoints);

            int sizeM = matchPoints.size();
            CvPoint center;
            CvSize imgSize;
            imgSize = cvGetSize(inImg[nowID]);
            center.x = imgSize.height/2;
            center.y = imgSize.width/2;
            batchToCylinder(matchPoints,center,FOCAL,matchPointC);
            int sizeC = matchPointC.size();
            if(matchPointC.size()> 0)
                Myransac2(matchPointC,homo,1,count);
            cvmSet(bestHomo,0,2,cvmGet(homo,0,2));
            cvmSet(bestHomo,1,2,cvmGet(homo,1,2));

            NextID = (nowID + 1) % filenum;
            matchPoints.clear();
            matchPointC.clear();

        }
}

void findTargetHomo(IplImage * inImg[MaxPics],ANNkd_tree * kdTree[MaxPics],vector < Feature >
keypoints[MaxPics],int nowID,int NextID,CvMat * bestHomo)
{
    vector < Sample > matchPoints;
    vector < SampleC > matchPointC;
    int match[30000]; //2015-05-25 wjs change from 1000 to 30000
    double dist[30000];
    CvMat * homo;
    int count;
    homo = cvCreateMat(3,3,CV_32FC1);

        featureMatch(keypoints[nowID],keypoints[NextID],kdTree[nowID],match,dist);
        int size2;
        size2 = keypoints[NextID].size();
        pickMatchPoint(keypoints[nowID],keypoints[NextID],match,dist,size2,matchPoints);
```

```
        int sizeM = matchPoints.size();
        CvPoint center;
        CvSize imgSize;
        imgSize = cvGetSize(inImg[nowID]);
        center.x = imgSize.height/2;
        center.y = imgSize.width/2;
        batchToCylinder(matchPoints,center,FOCAL,matchPointC);
        int sizeC = matchPointC.size();
        if(matchPointC.size()> 0)
            Myransac2(matchPointC,homo,1,count);
        cvmSet(bestHomo,0,2,cvmGet(homo,0,2));
        cvmSet(bestHomo,1,2,cvmGet(homo,1,2));


        matchPoints.clear();
        matchPointC.clear();


}

void seqNextImage2(IplImage * inImg[MaxPics],ANNkd_tree * kdTree[MaxPics],vector< Feature >
keypoints[MaxPics],int nowID,int &NextID,int ind[MaxPics],CvMat * bestHomo)
{
    vector< Sample > matchPoints;
    vector< SampleC > matchPointC;
    int match[30000];//2015-05-25 wjs change from 1000 to 30000
    double dist[30000];
    CvMat * homo;
    homo = cvCreateMat(3,3,CV_32FC1);
    int count;

    if(ind[nowID] == -1)
    {
        NextID = -1;
    }
    else
    {
        featureMatch(keypoints[nowID],keypoints[ind[nowID]],kdTree[nowID],match,dist);
        int size2;
        size2 = keypoints[ind[nowID]].size();
            pickMatchPoint (keypoints [nowID], keypoints [ind [nowID]], match, dist, size2,
            matchPoints);
```

```
            int sizeM = matchPoints.size();
            CvPoint center;
            CvSize imgSize;
            imgSize = cvGetSize(inImg[nowID]);
            center.x = imgSize.height/2;
            center.y = imgSize.width/2;
            batchToCylinder(matchPoints,center,FOCAL,matchPointC);
            int sizeC = matchPointC.size();
            if(matchPointC.size()>0)
                Myransac2(matchPointC,homo,1,count);
            cvmSet(bestHomo,0,2,cvmGet(homo,0,2));
            cvmSet(bestHomo,1,2,cvmGet(homo,1,2));

            NextID = ind[nowID];
            matchPoints.clear();
            matchPointC.clear();

        }
    }
```

以上源码展示的是进行 SIFT 特征提取及 ANN 网络分类的整个过程。

10.3 三张合成

由于水稻比较长，采用 X-ray 设备进行三次图像采集后，再进行图像的拼接，就可以获得一张完整的水稻图像，如图 10-4、图 10-5 和图 10-6 所示，三张合成后的图像如图 10-7 所示。

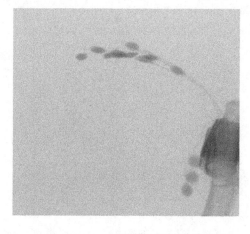

图 10-4 水稻左边 X 光图

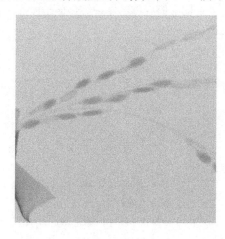

图 10-5 水稻右边 X 光图

| 第 10 章 | SIFT 和 ANN 实现任意顺序图像的合成

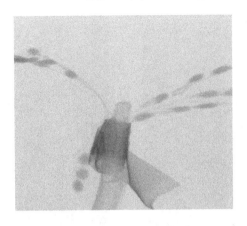

图 10-6　水稻下面 X 光图

图 10-7　三张水稻 X 光合成图片

10.4　五张合成

　　首先进行图像采集，采集五张公园图像，然后进行合成。获得的五张图像如图 10-8、图 10-9、图 10-10、图 10-11 和图 10-12 所示，合成后的结果如图 10-13 所示。

图 10-8　公园图像 1

图 10-9　公园图像 2

图 10-10　公园图像 3

图 10-11　公园图像 4

图 10-12　公园图像 5

图 10-13　五张公园图像合成图

10.5 七张合成

首先采集七张风景图(如图 10-14~图 10-20 所示),然后将有石头和青草,以及太阳的照射和阴影的多个图像进行合成,合成后的结果如图 10-21 所示。

图 10-14　风景图像 1

图 10-15　风景图像 2

图 10-16　风景图像 3

图 10-17　风景图像 4

图 10-18　风景图像 5

图 10-19　风景图像 6

图 10-20　风景图像 7

图 10-21　七张风景图像合成图

思 考 题

简答题

1. 图像合成的原理是什么？
2. 如果是无序地放入图像，图像合成是否可以自动识别顺序？

第 11 章

采用 Tesseract、BP、DP 进行 OCR 识别

11.1 字符识别方案选择

由于使用场景的不同,我们需要采用不同的 OCR 识别方案。大体上看,我们可以采用三种方法,即 Tesseract 方法,BP 神经网络方法和深度学习的方法。下面就这三种方案进行讲解和实现。

11.2 采用 Tesseract 识别英文和数字

采用 Tesseract 方法识别英文和数字,首先需要将 OpenCV 添加 Tesseract 模块后进行编译,生成相关的库,然后将头文件添加进去。代码如下所示。

```
#include "opencv2/text.hpp"
#include <opencv2/imgproc.hpp>
#include <opencv2/text/ocr.hpp>
```

首先读取一张图片,然后进行灰度化,接着进行二值化和黑白反色,最后识别。代码如下所示。

```
Mat src = imread("E:\\pngs\\test4.png");
    cv::Mat gray_src;
    if(src.channels() == 3)
    {
        cv::cvtColor(src, gray_src, CV_BGR2GRAY);
    }
    else
    {
        src.copyTo(gray_src);
    }
    Mat dst;
    cv::threshold(gray_src, dst, 0, 255, CV_THRESH_BINARY | CV_THRESH_OTSU);//灰度图像二值化
    cv::imshow("dst", dst);
```

```cpp
    cv::waitKey(0);

    cv::Mat not_image;
    cv::bitwise_not(dst, not_image);
    cv::imshow("not_image", not_image);
    cv::waitKey(0);
    //将切割出来的图片输入 tesseract 中
    const char * datapath = "E:\\mytessdata";
    const char * language = "eng";
    auto ocr = cv::text::OCRTesseract::create(datapath, language, NULL, cv::text::OEM_DEFAULT, cv::text::PSM_AUTO);
    auto output_text = ocr->run(not_image, 0, 0);
    std::cout <<"output_text = "<< output_text << std::endl;
```

准备识别的测试图片如图 11-1 所示,测试结果如图 11-2 所示。

图 11-1　准备识别的测试图片

图 11-2　采用 Tesseract 识别的结果

11.3　采用 BackPropagation 识别喷码字符

在深度学习方法应用之前,反向传播算法(BackPropagation,BP)在人工智能领域被成功地应用于车牌识别。BP 神经网络的应用主要分为三个部分,第一个部分是样本的准备,第二个部分是网络的训练,第三个部分是模型的推理。

第 11 章　采用 Tesseract、BP、DP 进行 OCR 识别

数字 0 的建模样例图片如图 11-3 所示。

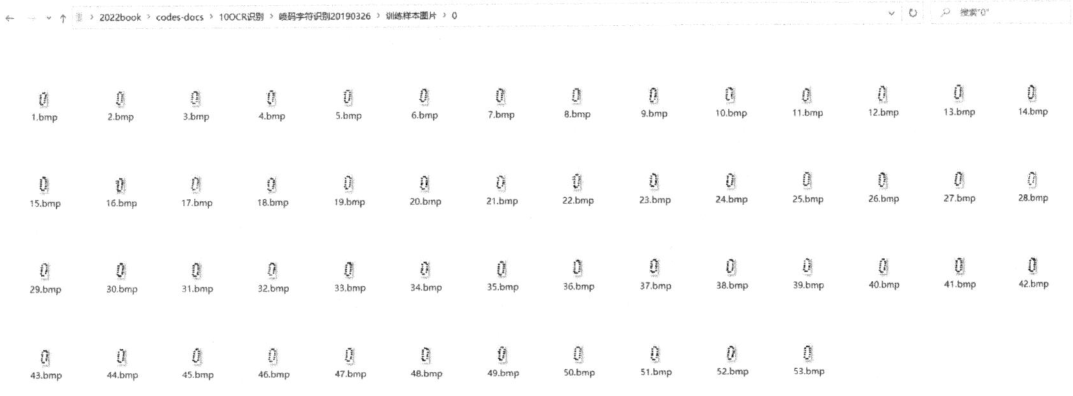

图 11-3　数字 0 的建模样例图片

模型的构建与训练的关键代码如下所示。

```
cout<<"training start...."<<endl;
Ptr<ANN_MLP> bp = ANN_MLP::create();
Mat layerSizes = (Mat_<int>(1, 5)<< image_rows * image_cols, int(image_rows * image_cols / 2),
    int(image_rows * image_cols / 4), int(image_rows * image_cols / 8), class_mun);
bp->setLayerSizes(layerSizes);
bp->setTrainMethod(ml::ANN_MLP::BACKPROP, 0.001,0.1);
bp->setActivationFunction(ANN_MLP::SIGMOID_SYM,1.0,1.0);
bp->setTermCriteria(TermCriteria(TermCriteria::MAX_ITER | TermCriteria::EPS, 10000, 0.0001));
bool trained = bp->train(trainingDataMat, ml::ROW_SAMPLE, labelsMat);
bp->save("MLPModel.xml");
cout<<"training finish...bpModel.xml saved "<<endl;
```

采用 BP 神经网络喷码识别的结果如图 11-4 所示。

图 11-4　采用 BP 神经网络喷码识别的效果

11.4 采用DeepLeraning识别中文手写体

考试试卷的自动化阅卷,已经是一个比较大的发展趋势。中文的相关信息,英文的相关信息,以及将纸质试卷,基于工业相机进行采集图像;然后进行图像的矫正;接着,图像算法人为地标注并生成一张Mask图片,提取BigROI;然后通过人工标注的小区域,进行小区域分割;接着通过Inception分类网络进行分类(中文,英文,符合);最后通过三个识别器,来进行识别。基于深度学习的中文手写识别器,基于OCRTesseract进行英文和数字识别,基于深度学习的符合识别器进行识别。

Deep Convolutional Network for Handwritten Chinese Character Recognition

Yuhao Zhang
Computer Science Department
Stanford University
zyh@stanford.edu

Abstract

In this project we explored the performance of deep convolutional neural network on recognizing handwritten Chinese characters. We ran experiments on a 200-class and a 3755-class dataset using convolutional networks with different depth and filter numbers. Experimental results show that deeper network with larger filter numbers give better test accuracy. We also provide a visualization of the learned network on the handwritten Chinese characters.

1. Introduction

Deep convolutional neural network (CNN) has become the architecture of choice for complex vision recognition problems for several years. There has been a lot of research on using deep CNN to recognize handwritten digits, English alphabets, or the more general Latin alphabets. Experiments have shown that well-constructed deep CNNs are powerful tools to tackle these challenges. As the recognition of characters in various languages has attracted much attention in the research community, a natural question is: How does deep CNN perform for recognizing more complex handwritten characters? In this project, we will explore the power of deep CNN on the classification of handwritten Chinese characters.

Compared to the task of recognizing handwritten digits and English alphabets, the recognition of handwritten Chinese characters is a more challenging task due to various reasons. Firstly, there are much more categories for Chi-

existence of joined-up handwriting makes the recognition even more difficult. For example, Figure 2 shows the influence of different handwriting styles on the appearance of handwritten Chinese characters. It is even a challenging task for a well-educated Chinese to recognize all the handwritten characters correctly.

In this project, we will focus on two specific questions: 1) How will the architecture and depth influence the accuracy of CNN on recognizing handwritten Chinese characters? 2) Does the extracted features make sense in terms of visualization? The rest of the report is organized as follows. We will first introduced the dataset and our network configurations in Section 2 and Section 3. Then we will introduce how we implement and train our networks in Section 4. Afterwards we present our experimental results in Section 5 and analyze the results in Section 6. Finally, we will discuss the related work in Section 7.

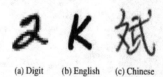

(a) Digit   (b) English   (c) Chinese

Figure 1: Example of handwritten characters

2. Data
2.1. Dataset

图 11-5 深度学习识别中文的文章

深度学习的模型构建关键代码来自如图11-5的文章,其对应的核心代码如下所示。

```python
def cnn():
    # (1-keep_prob) equals to dropout rate on fully-connected layers
    keep_prob = tf.placeholder(dtype = tf.float32, shape = [], name = 'keep_prob')
    # Set up places for data and label, so that we can feed data into network later
    img = tf.placeholder(tf.float32, shape = [None, 64, 64, 1], name = "img_batch")
    labels = tf.placeholder(tf.int64, shape = [None], name = "label_batch")
    # Structure references to: http://yuhao.im/files/Zhang_CNNChar.pdf,
    # however I adjust a little bit due to limited computational resource.
    # Four convolutional layers with kernel size of [3,3], and ReLu as activation function
    conv1 = slim.conv2d(img, 64, [3, 3], 1, padding = "SAME", scope = "conv1")
    pool1 = slim.max_pool2d(conv1, [2, 2], [2, 2], padding = "SAME")
    conv2 = slim.conv2d(pool1, 128, [3, 3], padding = "SAME", scope = "conv2")
    pool2 = slim.max_pool2d(conv2, [2, 2], [2, 2], padding = "SAME")
    conv3 = slim.conv2d(pool2, 256, [3, 3], padding = "SAME", scope = "conv3")
    pool3 = slim.max_pool2d(conv3, [2, 2], [2, 2], padding = "SAME")
    conv4 = slim.conv2d(pool3, 512, [3, 3], [2, 2], scope = "conv4", padding = "SAME")
    pool4 = slim.max_pool2d(conv4, [2, 2], [2, 2], padding = "SAME")
    # Flat the feature map so that we can connect it to fully-connected layers
    flat = slim.flatten(pool4)
    # Two fully-connected layers with dropout rate as mentioned at the start
    # First layer used tanh() as activation function
    fcnet1 = slim.fully_connected(slim.dropout(flat, keep_prob = keep_prob), 1024, activation
    _fn = tf.nn.tanh,
    scope = "fcnet1")
    fcnet2 = slim.fully_connected(slim.dropout(fcnet1, keep_prob = keep_prob), 3755,
    activation_fn = None, scope = "fcnet2")
    # loss function is defined as cross entropy on result of softmax function on last layer
    loss = tf.reduce_mean(tf.nn.sparse_softmax_cross_entropy_with_logits(logits = fcnet2,
    labels = labels))
    # compare result to actual label to get accuracy
    accuracy = tf.reduce_mean(tf.cast(tf.equal(tf.argmax(fcnet2, 1), labels), tf.float32))
    step = tf.get_variable("step", shape = [], initializer = tf.constant_initializer(0),
    trainable = False)
    # learning rate with exponential decay
    lrate = tf.train.exponential_decay(2e-4, step, decay_rate = 0.97, decay_steps = 2000,
    staircase = True)
    # Adam optimizer to decrease loss value
    optimizer = tf.train.AdamOptimizer(learning_rate = lrate).minimize(loss, global_step = step)
    prob_dist = tf.nn.softmax(fcnet2)
    val_top3, index_top3 = tf.nn.top_k(prob_dist, 3)
    # Write log into TensorBoard
```

```python
        tf.summary.scalar("loss", loss)
        tf.summary.scalar("accuracy", accuracy)
        summary = tf.summary.merge_all()
        return {"img": img,
                "label": labels,
                "global_step": step,
                "optimizer": optimizer,
                "loss": loss,
                "accuracy": accuracy,
                "summary": summary,
                'keep_prob': keep_prob,
                "val_top3": val_top3,
                "index_top3": index_top3
                }
```

深度学习中文识别网络架构如图 11-6 所示。

M5	M6-	M6	M6+	M7-1	M7-2	M9	M11
5 weight layers	6 weight layers	6 weight layers	6 weight layers	7 weight layers	7 weight layers	9 weight layers	11 weight layers
CNN Configurations							
input data (64 x 64 gray-scale image)							
conv3-64	conv3-32	conv3-64	conv3-80	conv3-64	conv3-64	conv3-64	conv3-64 conv3-64
maxpool							
conv3-128	conv3-64	conv3-128	conv3-160	conv3-128	conv3-128	conv3-128	conv3-128 conv3-128
maxpool							
conv3-256	conv3-128	conv3-256	conv3-320	conv3-256	conv3-256	conv3-256 conv3-256	conv3-256 conv3-256
maxpool							
	conv3-256	conv3-512	conv3-640	conv3-512 conv3-512	conv3-512	conv3-512 conv3-512	conv3-512 conv3-512
maxpool							
						FC-1024	
FC-1024							
FC-200 / FC-3755							
softmax							

图 11-6 深度学习中文识别网络架构

对选择的 3 755 个标准手写样本进行训练。代码如下所示。

```python
def train():
    with tf.Session() as sess:
        print("Start reading data")
        # Get batch tensor of data
        trn_imgs, trn_labels = batch(FLAGS.train_dir, FLAGS.batch_size, prepocess = True)
```

```python
    tst_imgs, tst_labels = batch(FLAGS.test_dir, FLAGS.batch_size)
    graph = cnn()
    # Preparation before training
    sess.run(tf.global_variables_initializer())
    coord = tf.train.Coordinator()
    threads = tf.train.start_queue_runners(sess, coord)
    saver = tf.train.Saver()
    if not os.path.isdir(FLAGS.logger_dir):
        os.mkdir(FLAGS.logger_dir)
    trn_summary = tf.summary.FileWriter(os.path.join(FLAGS.logger_dir, 'trn'), sess.graph)
    tst_summary = tf.summary.FileWriter(os.path.join(FLAGS.logger_dir, 'tst'))
    step = 0
    # If received restore flag, train from last checkpoint
    if FLAGS.restore:
        # Get last checkpoint in checkpoint directory
        checkpoint = tf.train.latest_checkpoint(FLAGS.checkpoint)
        if checkpoint:
            # Restore data from checkpoint
            saver.restore(sess, checkpoint)
            step += int(checkpoint.split('-')[-1])
    print("Train from checkpoint")
    print("Start training")
        while not coord.should_stop():
        # Get actual data
        trn_img_batch, trn_label_batch = sess.run([trn_imgs, trn_labels])
        # Prepare parameters for network
        graph_dict = {graph['img']: trn_img_batch, graph['label']: trn_label_batch, graph
        ['keep_prob']: 0.8}
        # Feed and parameter into network
        opt, loss, summary, step = sess.run(
            [graph['optimizer'], graph['loss'], graph['summary'], graph['global_step']],
            feed_dict = graph_dict)
        trn_summary.add_summary(summary, step)
        print("# " + str(step) + " with loss " + str(loss))
        if step > FLAGS.max_step:
            break
        # Evaluate current network based on test dataset
        if(step % 500 == 0) and (step >= 500):
            tst_img_batch, tst_label_batch = sess.run([tst_imgs, tst_labels])
            graph_dict = {graph['img']: tst_img_batch, graph['label']: tst_label_batch,
            graph['keep_prob']: 1.0}
```

```python
            accuracy, test_summary = sess.run([graph['accuracy'], graph['summary']], feed
            _dict = graph_dict)
            tst_summary.add_summary(test_summary, step)
            print("Accuracy: %.8f" % accuracy)
            # Save checkpoint
            if step % 10000 == 0:
                saver.save(sess, os.path.join(FLAGS.checkpoint, 'hccr'), global_step =
                graph['global_step'])
        coord.join(threads)
        saver.save(sess, os.path.join(FLAGS.checkpoint, 'hccr'), global_step = graph['global_step'])
        sess.close()
    return
```

将训练好的模型进行测试。代码如下所示。

```python
def test(path):
    # Read test picture and resize it, turn it to grey scale.
    tst_image = cv2.imread(path, cv2.IMREAD_GRAYSCALE)
    tst_image = cv2.resize(tst_image, (64, 64))
    tst_image = numpy.asarray(tst_image) / 255.0
    tst_image = tst_image.reshape([-1, 64, 64, 1])
    # feed the test picture into network and estimate probability distribution
    with tf.Session() as sess:
        graph = cnn()
        saver = tf.train.Saver()
        saver.restore(sess = sess, save_path = tf.train.latest_checkpoint(FLAGS.checkpoint))
        graph_dict = {graph['img']: tst_image, graph['keep_prob']: 1.0}
        val, index = sess.run([graph['val_top3'], graph['index_top3']], feed_dict = graph_dict)
        for i in range(3):
            print("Probability: %.5f" % val[0][i] + " with label:" + str(index[0][i]))
        path = FLAGS.train_dir + "/" + '%0.5d' % index[0][0]
        # select one of the picture from the label with top 1 probability
        for root, dir, files in os.walk(path):
            for f in files:
                img = cv2.imread(path + "/" + f)
                enlarged = cv2.resize(img, (img.shape[1] * 5, img.shape[0] * 5))
                cv2.imshow("Top1", enlarged)
                cv2.waitKey()
                break
            break
    return val, index
```

至此，三种 OCR 识别方法介绍完毕，大家可以根据具体的应用选择不同的实现方法。

思 考 题

简答题

1. 用 Tesseract 方法进行 OCR 识别时,如何对原始图像进行处理?
2. 喷码字符识别中采用的反向传播 BP 网络的模型层数最多可以设置多少层?
3. 深度学习进行中文识别时,需要将中文文字进行分割,请列举深度学习的分割算法。

第 12 章
Python 和 Tensorflow 深度学习分类网络

12.1 分类网络

本章通过菜品分类的例子来学习深度学习分类网络。深度学习的卷积神经网络(CNN)和循环神经网络(RNN)是两个代表。本章重点学习分类网络 InceptionV3,它属于 CNN 网络之一。总体来看,菜品分类可以分为以下步骤。

第一步,深度学习编译环境。如 Win 10,CUDA 10.0,CUDNN 7.0,Python 3.7.5,Tensorflow 1.15.2,Keras 2.3.0 等。

第二步,数据的准备与转换。将数据从 jpg/png/bmp 转换为 Tensorflow 能识别的数据格式.tfrecord。

第三步,模型的构建与训练。采用 InceptionV3 网络进行训练。

第四步,模型冻结为 PB 和测试。训练代码如下所示。

```
python train_image_classifier.py --train_dir=satellite/train_dir --dataset_name=satellite --dataset_split_name=train --dataset_dir=satellite/data --model_name=inception_v3 --checkpoint_path=satellite/pretrained/inception_v3.ckpt --checkpoint_exclude_scopes=InceptionV3/Logits,InceptionV3/AuxLogits --trainable_scopes=InceptionV3/Logits,InceptionV3/AuxLogits --max_number_of_steps=100000 --batch_size=2 --learning_rate=0.001 --learning_rate_decay_type=fixed --save_interval_secs=300 --save_summaries_secs=20 --log_every_n_steps=10 --optimizer=rmsprop --weight_decay=0.00004

导出代码第一步:python export_inference_graph.py --alsologtostderr --model_name=inception_v3 --output_file=satellite/inception_v3_inf_graph.pb --dataset_name satellite

导出代码第二步:python freeze_graph.py --input_graph slim/satellite/inception_v3_inf_graph.pb --input_checkpoint slim/satellite/train_dir/model.ckpt-100000 --input_binary true --output_node_names InceptionV3/Predictions/Reshape_1 --output_graph slim/satellite/frozen_graph.pb
```

第五步,C++ 与 OpenCV 高速推理执行。

12.2 模型网络

菜品分类的网络模型可以分为以下步骤。

第一步，在学校或单位的食堂，用手机拍四种菜品照片，每种菜品拍100张，然后统一用jpg格式存储。

第二步，将这批数据的80%进行训练，20%进行测试。那么，训练的数据总体样本为80×4，测试的总体样本为20×4，并将这批数据转换为.tfrecord。

第三步，采用迁移学习的方式，下载InceptionV3的原始网络模型，用菜品的训练样本，完成模型的训练后，生成新的优化模型，进行冻结为PB，进行测试。

12.3 推理执行

在实际工程环境中，大多数项目采用Python完成模型的训练，采用C++完成模型的高效推理与执行，这样既可以将训练与执行分开来开发，又可以让离线训练与在线执行各自发挥自己的优势。对于C++版本的OpenCV在深度学习的重点是完成模型的高效推理。

以下是OpenCV在分类网络模型推理的应用，读出分类的名称的详细代码如下所示。

```cpp
vector<String> readClassNames()
{
    vector<String> classNames;
    fstream fp(labels_txt_file);
    if(!fp.is_open())
    {
        cout <<"does not open"<< endl;
        exit(-1);
    }
    string name;
    while(!fp.eof())
    {
        getline(fp, name);
        if(name.length())
            classNames.push_back(name);
    }
    fp.close();
    return classNames;
}
```

模型的装载代码如下所示。

```cpp
Net net = readNetFromTensorflow(tf_pb_file);
```

模型的推理代码如下所示。

```cpp
const cv::Scalar mean_temp = cv::Scalar(B, G, R);
//Mat inputBlob = blobFromImage(src, 0.00390625f, Size(w, h), mean_temp, false, false);
Mat inputBlob = blobFromImage(src, 0.00390625f, Size(w, h), mean_temp, false, false);
```

```
//执行图像分类
Mat prob;
net.setInput(inputBlob, "input");
//prob = net.forward("softmax");
prob = net.forward();
cout << prob << endl;

//得到最大分类概率
Mat probMat = prob.reshape(1,1); Point classNumber;  double classProb;
minMaxLoc(probMat, NULL, &classProb, NULL, &classNumber);
//显示文本
putText(src,labels.at(classidx),Point(20,20),FONT_HERSHEY_SIMPLEX,1.0,Scalar(0,0,255),2,8);
imshow("Image Classfication", src);
waitKey(0);
```

通过上面的过程描述和关键代码介绍后,下面给出整个实验报告。

12.4 实验报告

实验项目:菜品识别

一、训练环境

1. Windows 10
2. Python 3.7.5
3. Tensorflow==1.15.2

二、实验目的

通过图像分类进行菜品的识别。

三、实验过程

首先准备好四种菜品,然后分别给四种菜品拍 100 张照片,其中每种菜品照片中的 80 张用于训练样本,20 张用于测试样本,并做好标签,放到 data_prepare/pic/train 和 data_prepare/pic/validation 中,如图 12-1 和图 12-2 所示。

在 data_prepare/下,执行命令如图 12-3 所示。

在 pic 文件中生成一个 label 文件和四个 tf-record 文件如图 12-4 所示。

在 slim 文件里新建 satellite 文件和在 satellite 文件下新建 data、pretrained 和 train_dir 三个文件,如图 12-5 所示。将转换生成的五个文件复制到 slim\satellite\data 下。如图 12-6 所示。

修改 slim\datasets\satellite.py 文件如图 12-7 和图 12-8 所示。

第12章 Python 和 Tensorflow 深度学习分类网络

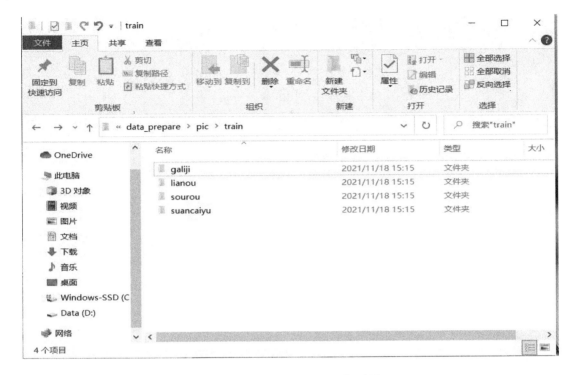

图 12-1 用于训练样本的四种菜品的标签

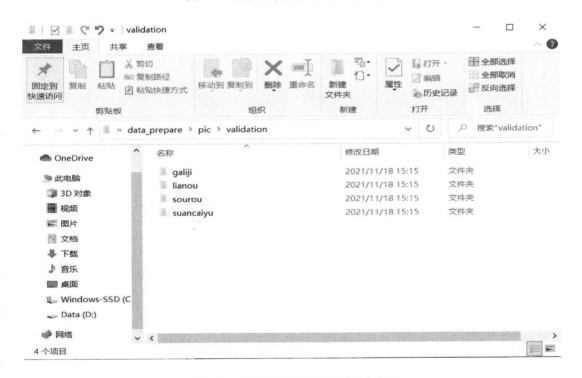

图 12-2 用于测试样本的四种菜品的标签

· 167 ·

```
C:\Windows\system32>D:

D:\>cd D:\Desktop\python3_tf_slim_image_classify-master\data_prepare

D:\Desktop\python_tf_slim_image_classify-master\data_prepare>python data_convert.py -t pic/ --0 --validation-shards 2 -
-num-threads 2 --dataset-name satellite
```

图 12-3 将数据转换为 Tensorflow 的输入格式

图 12-4 生成了一个 label 文件和四个 tf-record 文件

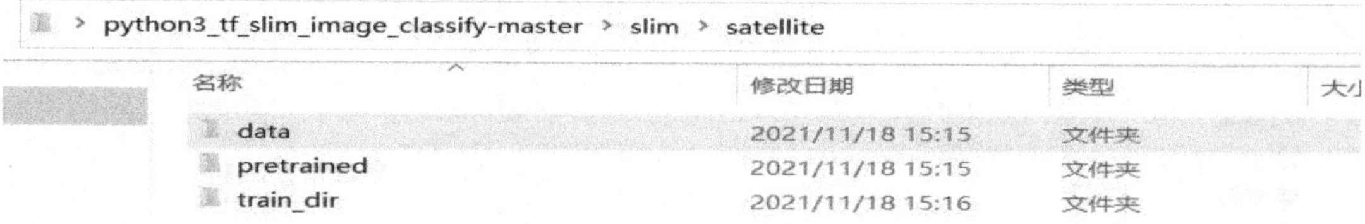

图 12-5 文件拷贝到对应的数据下面

图 12-6 对应的 data 下面的结果显示

```python
import ...

slim = tf.contrib.slim

_FILE_PATTERN = 'satellite_%s_*.tfrecord'

SPLITS_TO_SIZES = {'train': 320, 'validation': 80}

_NUM_CLASSES = 4

_ITEMS_TO_DESCRIPTIONS = {
    'image': 'A color image of varying size.',
    'label': 'A single integer between 0 and 4',
}


def get_split(split_name, dataset_dir, file_pattern=None, reader=None):
    """Gets a dataset tuple with instructions for reading flowers.
```

图 12-7 源文件 satellite.py 修改 1

```python
    if split_name not in SPLITS_TO_SIZES:
        raise ValueError('split name %s was not recognized.' % split_name)

    if not file_pattern:
        file_pattern = _FILE_PATTERN
    file_pattern = os.path.join(dataset_dir, file_pattern % split_name)

    # Allowing None in the signature so that dataset_factory can use the default.
    if reader is None:
        reader = tf.TFRecordReader

    keys_to_features = {
        'image/encoded': tf.FixedLenFeature((), tf.string, default_value=''),
        'image/format': tf.FixedLenFeature((), tf.string, default_value='jpg'),
        'image/class/label': tf.FixedLenFeature(
            [], tf.int64, default_value=tf.zeros([], dtype=tf.int64)),
    }
```

图 12-8 源文件 satellite.py 修改 2

（1）在 slim/ 文件夹下执行如下命令，进行 5 001 次训练。

python train_image_classifier.py - - train_dir = satellite/train_dir - - dataset_name = satellite - - dataset_split_name = train - - dataset_dir = satellite/data - - model_name = inception_ v3 --checkpoint_path=satellite/pretrained/inception_v3.ckpt --checkpoint_exclude_scopes=InceptionV3/Logits,InceptionV3/AuxLogits --trainable_scopes=InceptionV3/Logits,InceptionV3/AuxLogits --max_number_of_steps=5001 --batch_size=8 --learning_rate=0.001 --learning_rate_decay_type=fixed --save_interval_secs=300 --save_summaries_secs=2 --log_every_n_steps=10 --optimizer=rmsprop --weight_decay=0.00004

（2）在 slim/文件夹下执行如下命令，进行模型能力评估，结果如图 12-9 所示。

Python eval_image_classifier.py --checkpoint_path=satellite/train_dir --eval_dir=satellite/eval_dir --dataset_name=satellite --dataset_split_name=validation --dataset_dir=satellite/data --model_name=inception_v3

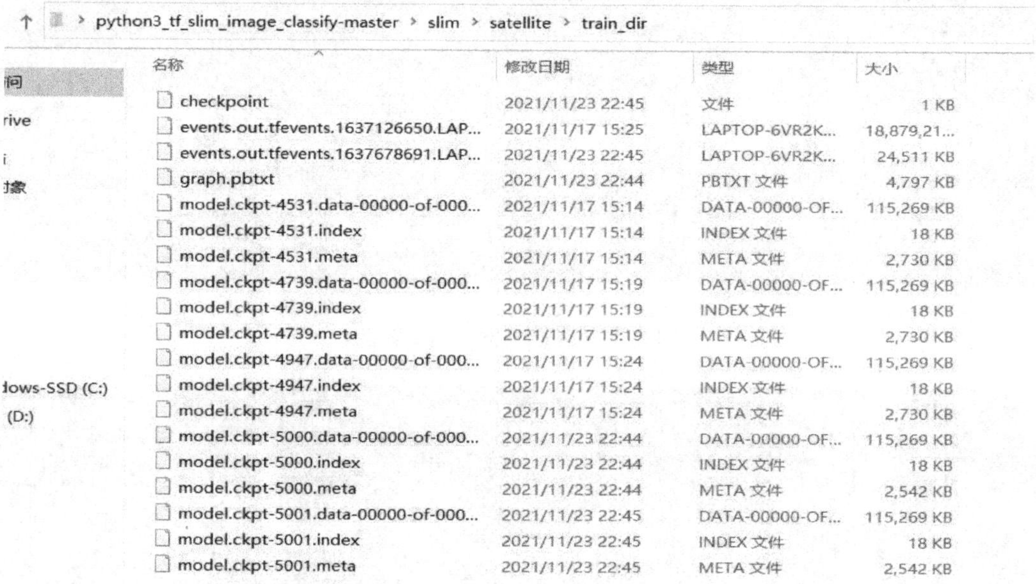

图 12-9 训练次数 5 000 后的结果

为了将训练好的模型部署到各个目标平台，需要将训练好的模型导出为标准格式文件。

（3）在 slim/文件夹下面执行如下命令。

python export_inference_graph.py --alsologtostderr --model_name=inception_v3 --output_file=satellite/inception_v3_inf_graph.pb --dataset_name satellite

（4）执行后会在 satellite 中生成 inception_v3_inf_graph.pb 文件如图 12-10 所示。

图 12-10 生成的 pb 文件

第 12 章 Python 和 Tensorflow 深度学习分类网络

在项目根目录执行如下命令，如图 12-11 所示。

python freeze_graph.py --input_graph slim/satellite/inception_v3_inf_graph.pb --input_checkpoint slim/satellite/train_dir/model.ckpt-5001 --input_binary true --output_node_names InceptionV3/Predictions/Reshape_1 --output_graph slim/satellite/frozen_graph.pb

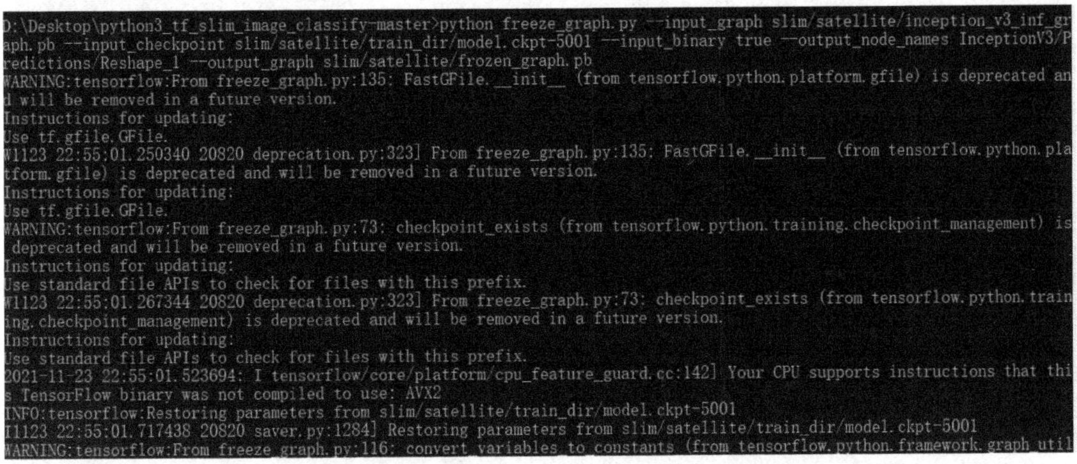

图 12-11　冻结生成的最后的 pb 文件的过程

执行后生成 frozen_graph.pb 文件操作，如图 12-12 所示。

图 12-12　生成网络模型的 pb 文件

测试的图片并且命名为 test_image 保存，如图 12-13 所示。

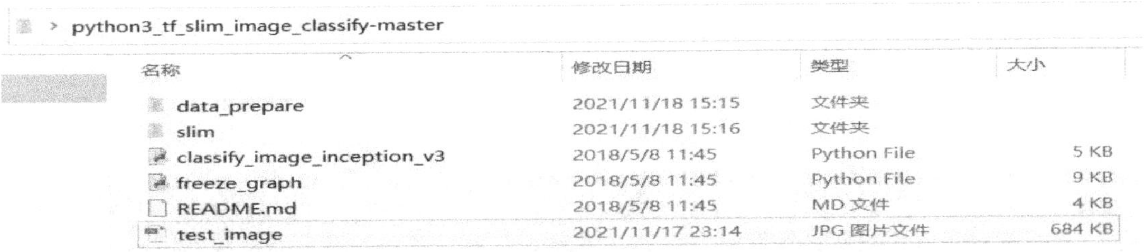

图 12-13　测试图片的准备

对单张图片进行预测在项目根目录执行如下命令,得出的结果如图 12-14 所示。

```
python classify_image_inception_v3.py --model_path slim/satellite/frozen_graph.pb --label_path data_prepare/pic/label.txt --image_file test_image.jpg
```

```
W1123 22:58:25.766924 19452 deprecation.py:323] From classify_image_inception_v3.py:85: FastGFile.__init__ (from tensorflow.python.platform.gfile) is deprecated and will be removed in a future version.
Instructions for updating:
Use tf.gfile.GFile.
WARNING:tensorflow:From classify_image_inception_v3.py:70: The name tf.image.resize_bilinear is deprecated. Please use tf.compat.v1.image.resize_bilinear instead.

W1123 22:58:26.121381 19452 module_wrapper.py:139] From classify_image_inception_v3.py:70: The name tf.image.resize_bilinear is deprecated. Please use tf.compat.v1.image.resize_bilinear instead.

WARNING:tensorflow:From classify_image_inception_v3.py:89: The name tf.Session is deprecated. Please use tf.compat.v1.Session instead.

W1123 22:58:26.127403 19452 module_wrapper.py:139] From classify_image_inception_v3.py:89: The name tf.Session is deprecated. Please use tf.compat.v1.Session instead.

2021-11-23 22:58:26.128855: I tensorflow/core/platform/cpu_feature_guard.cc:142] Your CPU supports instructions that this TensorFlow binary was not compiled to use: AVX2
WARNING:tensorflow:From classify_image_inception_v3.py:53: The name tf.GraphDef is deprecated. Please use tf.compat.v1.GraphDef instead.

W1123 22:58:26.235511 19452 module_wrapper.py:139] From classify_image_inception_v3.py:53: The name tf.GraphDef is deprecated. Please use tf.compat.v1.GraphDef instead.

suancaiyu (score = 6.34915)
sourou (score = -1.64911)
galiji (score = -1.96281)
lianou (score = -2.80678)

D:\Desktop\python3_tf_slim_image_classify-master>
```

图 12-14　深度学习中文识别网格架构

四、实验总结

其他一组数据的实验结果如图 12-15 所示。

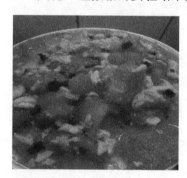

```
images/water_onion.jpg
lakeside, lakeshore (score = 0.32870)
corn (score = 0.19709)
ear, spike, capitulum (score = 0.15878)
yellow lady's slipper, yellow lady-slipper, Cypripedium calceolus, Cypripedium parviflorum (score = 0.01670)
hay (score = 0.01296)
```

图 12-15　深度学习分类测试效果展示

第 12 章 Python 和 Tensorflow 深度学习分类网络

整体识别还不错,准确率达到 99%,测试里面有几张酸痛咕噜肉识别成西红柿炒蛋,可能是因为样本训练不够充分。

思 考 题

简答题

1. 深度学习中的分类器有哪些?并说明各自的优缺点。
2. Python 训练模型后,为什么需要 C++执行?
3. Tensorflow 的版本为什么选择 1.15.2 版本?如果选择 2.6.0 版本,如何完成分类的功能?

第 13 章
Python 和 Tensorflow 深度学习分割网络

13.1　图像标注

图像标注软件 labelme,是在安装完 Python 后,采用在线式安装方式安装。推荐使用 Python 3.7.5 的版本,因为后面 Tensorflow 的版本是 1.15.2,所以如果 Python 版本过高,会出现一些 Python 语法与版本不兼容的问题。

在安装 labelme 时,如果 pip 不是最新版本,需要将 pip 进行升级,升级方法如图 13-1 所示,具体操作如下所示。

图 13-1　升级 pip 后安装 labelme 的方法

```
python.exe-m pip install-upgrade pip
```

安装 labelme 的方法：pip install labelme。

在 labelme 常用图标中，一般选择打开文件夹的方式，labelme 常用图标的功能如图 13-2 所示。

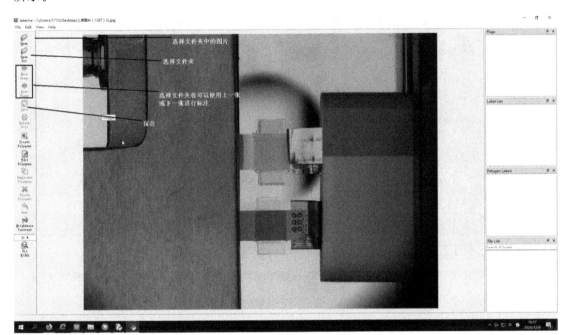

图 13-2　labelme 常用图标的功能

选择使用多边形的区域标注方法，如图 13-3 所示。

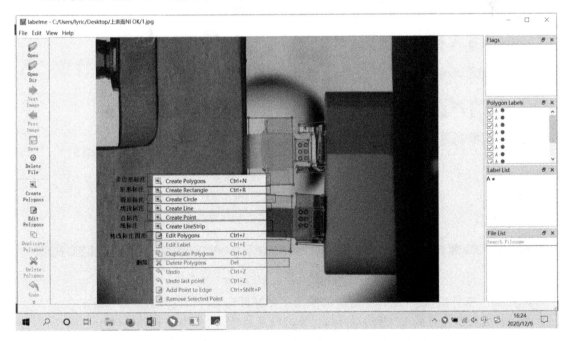

图 13-3　多边形的区域标注方法

选择使用多边形标注方法后,当一个闭合区域完成后,会让人对区域进行命名。这时,我们推荐用 A,B,C 等大写的单个字母来命名,这样做的好处在于解析时可以按照名字来解析。将所需目标区域标注好后进行命名,推荐使用名称 A,具体操作如图 13-4 所示。

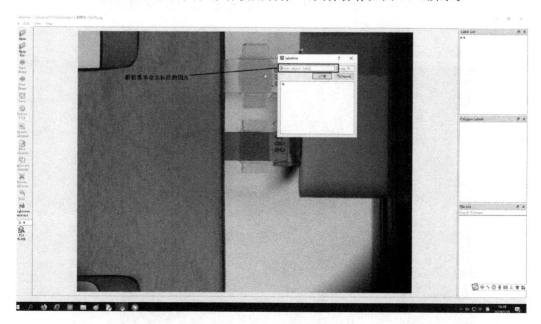

图 13-4　输入闭合区域的名称:A

保存图片后自动生成 json 文件,如图 13-5 所示。

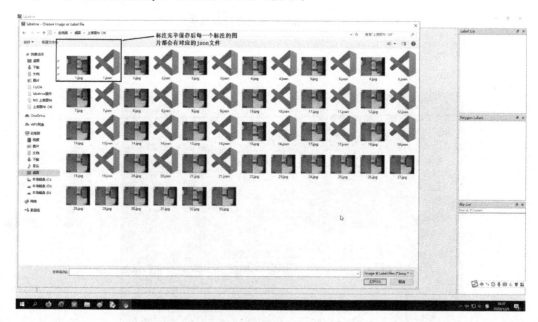

图 13-5　保存图片后自动生成 json 文件

解析 json 文件后,就有了一一对应的原始文件和目标文件,具体如图 13-6 所示。

通过将大量标注好的图片给人工智能"学习",经过训练,模型能自动提取出目标区域,标注时应标好,因为标注的过程相当于向计算机授课,只有老师教对了,学生才能学会。

第 13 章 Python 和 Tensorflow 深度学习分割网络

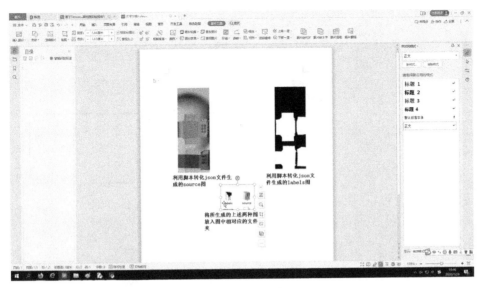

图 13-6 原图和解析 json 文件后生成的图像

13.2 图像提取

提取出所需的大区域只是相当于拿到了一张试卷,接下来要做的就是把它分给不同的老师批阅。这个时候就要进行图像分割。使用 Microsoft Visual Studio 运行脚本,依次填入新原图路径和结果图路径,以及被分成 10 张图之后各自的存放路径,如图 13-7 所示。

图 13-7 C++解析 json 的源码

10 个区域分别被放进不同的文件夹,如图 13-8 所示。

注意事项:
1. 将样本分为训练样本和测试样本。
2. 在训练过程中,不要中断,否则会重新训练。

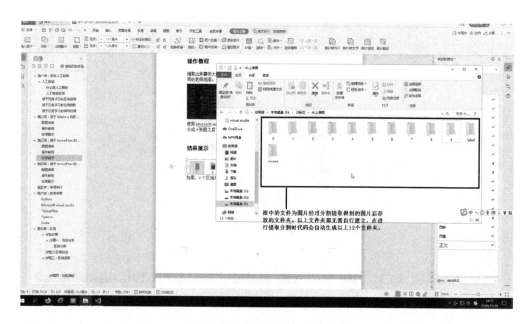

图 13-8　解析后生成的文件夹

如图 13-9 所示，图像是通过工业相机采集到的原始图像，下面的图像的白色区域就是缺陷的区域。大量的一一对应的原始图像和图像的缺陷区域为模型训练提供大数据素材。

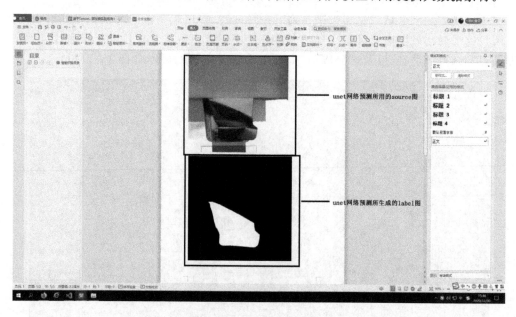

图 13-9　原图图像和解析 json 后生成的目标图像

13.3　模型训练

分类网络采用 Unet 网络来构建，网络设计采用下九层和上九层的模式。具体操作如图 13-10 所示，将整个数据集合的 80% 作为训练集合，20% 作为验证集合。

第 13 章　Python 和 Tensorflow 深度学习分割网络

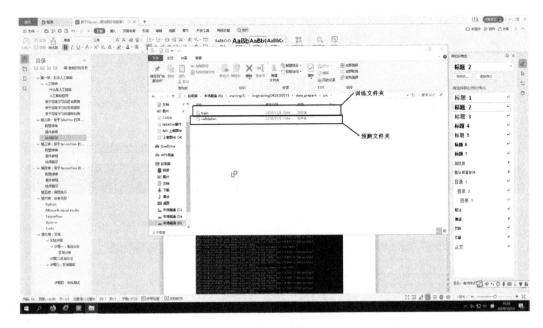

图 13-10　原始图像和目标图像放入 train 训练文件夹中

13.4　版本控制

因为语言的优势性，C++比 Python 的执行效率更高，因此配置 C++版的 Tensorflow 来推理执行深度学习模型更贴近项目对速度的需求。运行环境和硬件对应版本：Microsoft Visual Studio 2019＋OpenCV 4.5.3，GPU 与 CUDA 和 CUDNN 的版本说明如表 13-1 所示。

表 13-1　GPU 与 CUDA 和 CUDNN 的版本说明

GPU	CUDA 版本	CUDNN 版本
GTX 1050Ti(10 系)	10.1	7.6
GTX 16xx(16 系)	10.1	7.6
RTX 2080Ti(20 系)	10.1	7.6
RTX 3090(30 系)	11.2	8.1
RTX A6000	11.2	8.1

C 版 Tensorflow(根据 CUDA 和 CUDNN 版本选择安装)的版本说明如表 13-2 所示。

表 13-2　GPU 与 Tensorflow API 的版本说明

CUDA 版本	CUDNN 版本	Tensorflow API 版本
10.1	7.6	2.3.0
11.2	8.1	2.6.0

13.5 安装过程

以 CUDA 10.1＋CUDNN 7.6 为例。CUDA 下载地址：https://developer.nvidia.cn/cuda-toolkit-archive。下载页面如图 13-11 所示。

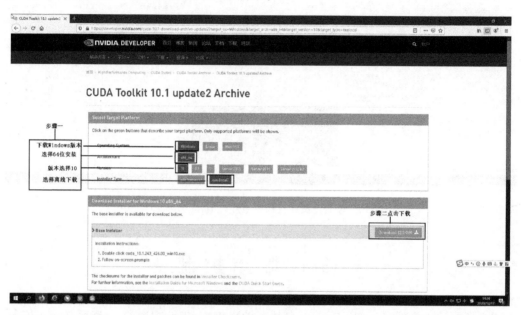

图 13-11　CUDA 10.1 官网下载页面

下载完毕后打开安装包，选择安装路径，如图 13-12 所示。

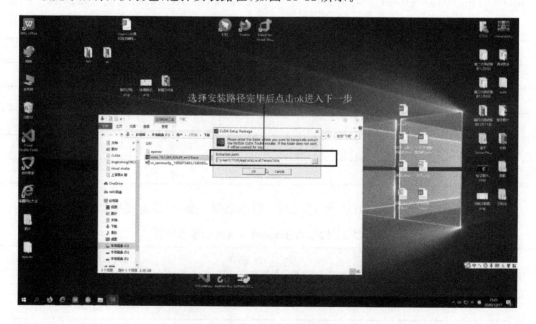

图 13-12　选择安装路径

安装进度条如图 13-13 所示。

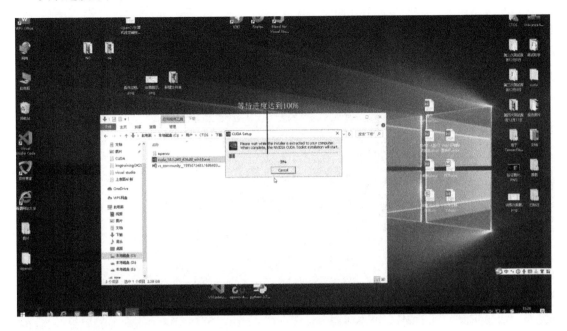

图 13-13　安装进度条

GPU 硬件环境检测如图 13-14 所示。

图 13-14　GPU 硬件环境检测

接受协议单击"同意"并继续,如图13-15所示。

图13-15 选择同意协议

勾选自定义安装并单击"下一步",如图13-16所示。

图13-16 勾选自定义安装

第 13 章 | Python 和 Tensorflow 深度学习分割网络

取消勾选选项,如图 13-17 和图 13-18 所示。

图 13-17 取消不用的勾选

图 13-18 取消此处的勾选项

单击"下一步",打开命令行验证安装是否完成,如图 13-19 所示。
输入"nvcc--version",如图 13-20 所示。

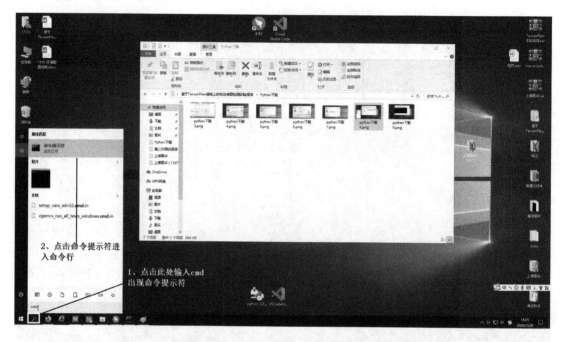

图 13-19 选择 Win 10 左下角第 2 个图标后点击命令提示符

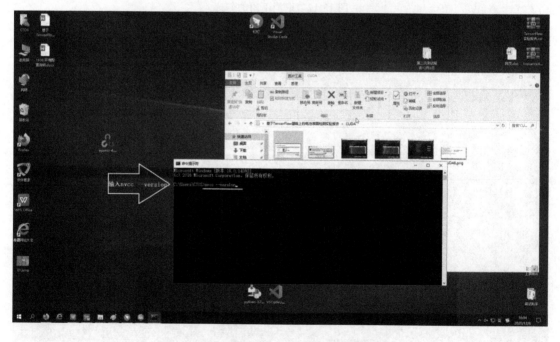

图 13-20 验证 GPU 的 CUDA 是否安装完成

输入"nvcc--version"后出现下图框中文字则进入下一步,继续输入"set cuda",然后按回车键,如图 13-21 所示。与图 13-22 所示一致则说明 CUDA 安装完毕。

第 13 章 | Python 和 Tensorflow 深度学习分割网络

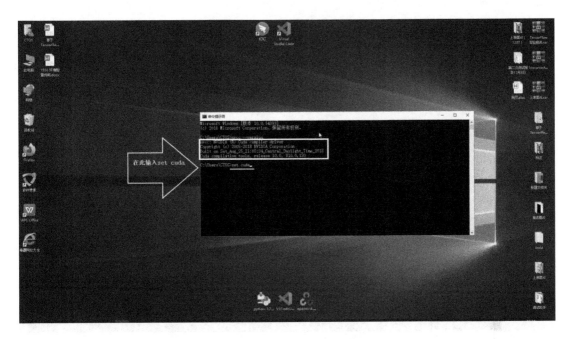

图 13-21 输入"set cuda"

图 13-22 安装 CUDA 后的相关路径

下载与 CUDA 版本对应的 CUDNN。例如，CUDA 10.1 对应的是 CUDNN 7.6，CUDNN 压缩包解压后与 CUDA 文件夹中的 bin、inciude、lib 三个文件夹进行替换即可，如图 13-23 所示。

若需下载不同的版本，改变上面链接的对应的版本号即可。下载后解压到 E 盘，如图 13-24 所示。

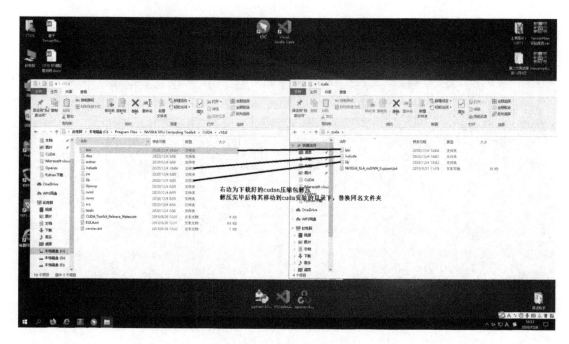

图 13-23　安装 CUDNN 的方法

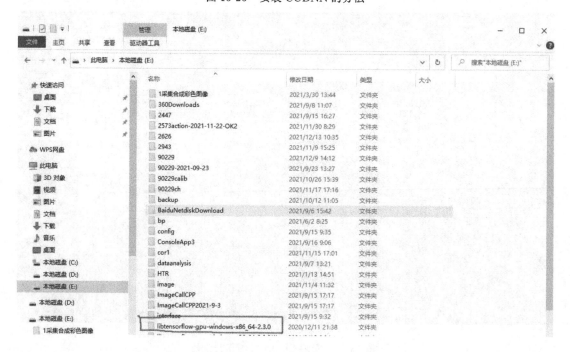

图 13-24　GPU 1050,2080ti 对应的 Tensorflow API 版本

网络资源参考如下。

（1）Tensorflow C++从训练到部署：环境搭建。

http://www.liuxiao.org/2018/08/ubuntu-tensorflow-c-%e4%bb%8e%e8%ae%ad%e7%bb%83%e5%88%b0%e9%a2%84%e6%b5%8b1%ef%bc%9a%e7%8e%af%e5%a2%83%e6%90%ad%e5%bb%ba/

(2) Tensorflow C++环境配置。

https：//www.jianshu.com/p/cbf0e668f135

(3) CUDNN 下载地址。

https：//developer.nvidia.cn/rdp/cudnn-archive

(4) C 版 Tensorflow 的安装下载地址。

https：//storage.googleapis.com/tensorflow/libtensorflow/libtensorflow-gpu-windows-x86_64-2.3.0.zip

思 考 题

简答题

1. labelme 标注软件是如何安装的？
2. 深度学习平台的 GPU 与 CUDA，CUDNN 的版本对应关系如何？
3. Unet 网络的基本原理和功能是什么？

第 14 章
C♯和 C++接口设计 CLR

14.1 接口 CLR

在自动化领域,大多数软件系统采用 C♯作为主系统开发语言,用来连接 PLC 和控制 CCD,以及完成整个人机交互。面向大多数采用 C♯作为 UI 设计的方法,从 C♯来讲,需要两个转换:Bitmap 转为 Mat 后进行 C++算法执行后,将结果 Mat 转为 Bitmap 给 C♯显示,整个开发流程如图 14-1 所示。

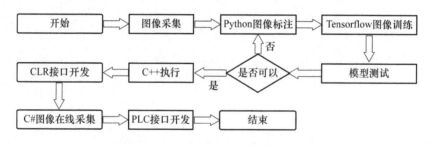

图 14-1 深度学习模型训练与 C♯平台框架

在非标定制自动化的设计当中,大多数企业会选择 C♯作为图像采集和 GUI 界面的开发,Python 作为图像标注和深度学习的开发语言,C++作为高效执行的语言,被大量采用。机器视觉常规开发如表 14-1 所示。

表 14-1 深度学习模型训练与 C♯平台框架

机器视觉方法	占比	开发语言	开发框架	开发阶段
打光方案	10%	Python	Pyqt5	第一阶段
图像前处理	80%	C++	OpenCV	第二阶段
图像后处理	5%	C++	OpenCV	第三阶段
在线执行	5%	C♯	CLR	第四阶段

深度学习阶段开发如表 14-2 所示。

表 14-2 深度学习开发阶段

深度学习方法	占比	开发语言	开发框架	开发阶段
打光方案	10%	Python	Pyqt5	第一阶段
模型训练	80%	Python	Tensorflow	第二阶段
图像后处理	5%	C++	OpenCV	第三阶段
接口执行	5%	C#	CLR	第四阶段

总体来看，C#作为机器视觉系统的 HMI（人机接口）主要的开发语言之一，大多数应用算法采用 C++来开发，所以 CLR 是将 C#与 C++连接起来的重要桥梁。

14.2 图像转换

如何将 C#转换为 C++的格式呢？在图像处理中，本质上就是 System∷Drawing∷Bitmap 转换为 C++的 cv∷Mat，具体实现代码如下所示。

```cpp
cv∷Mat BitmapToMat(System∷Drawing∷Bitmap^ bitmap)
{
    IplImage * tmp;

    System∷Drawing∷Imaging∷BitmapData^ bmData = bitmap->LockBits(System∷Drawing∷
    Rectangle(0, 0, bitmap->Width, bitmap->Height), System∷Drawing∷Imaging∷
    ImageLockMode∷ReadOnly, bitmap->PixelFormat);
    if(bitmap->PixelFormat == System∷Drawing∷Imaging∷PixelFormat∷Format8bppIndexed)
    {
        tmp = cvCreateImage(cvSize(bitmap->Width, bitmap->Height), IPL_DEPTH_8U, 1);
        tmp->imageData = (char *)bmData->Scan0.ToPointer();
    }
    else if(bitmap->PixelFormat == System∷Drawing∷Imaging∷PixelFormat∷
    Format24bppRgb)
    {
        tmp = cvCreateImage(cvSize(bitmap->Width, bitmap->Height), IPL_DEPTH_8U, 3);
        tmp->imageData = (char *)bmData->Scan0.ToPointer();
    }
    else
    {
        tmp = cvCreateImage(cvSize(bitmap->Width, bitmap->Height), IPL_DEPTH_8U, 4);
        tmp->imageData = (char *)bmData->Scan0.ToPointer();
    }

    bitmap->UnlockBits(bmData);
    cv∷Mat cvImage = cv∷cvarrToMat(tmp);
```

```cpp
    cv::Mat mv[4];
    cv::split(cvImage, mv);
    /*cv::imshow("1", mv[0]);
    cv::imshow("2", mv[1]);
    cv::imshow("3", mv[2]);
    cv::imshow("4", mv[3]);*/
    cv::Mat result;
    cv::merge(mv, 3, result);
    cvReleaseImage(&tmp);
    return result;
}
```

C++的 Mat 转 C#的 Bitmap，实现代码如下所示。

```cpp
//////////////////////////////////////2nd Functions//////////////////////////////////////
/*---------------------------
* 功能：将图像格式由 cv::Mat 转换为 System::Drawing::Bitmap
*      -不拷贝图像数据
*---------------------------
* 函数：ConvertMatToBitmap
* 访问：public
* 返回：Bitmap 图像指针，若转换失败，则返回的图像宽高均为1
*
* 参数：cvImg        [in]    OpenCV 图像
*/
System::Drawing::Bitmap^ ConvertMatToBitmap(cv::Mat& cvImg)
{
    System::Drawing::Bitmap^ bmpImg;
    if(cvImg.depth() != CV_8U)
    {
        bmpImg = gcnew System::Drawing::Bitmap(1, 1, System::Drawing::Imaging::
            PixelFormat::Format8bppIndexed);
        return(bmpImg);
    }
    //彩色
    if(cvImg.channels() == 3)
    {
        bmpImg = gcnewBitmap(
        cvImg.cols,
        cvImg.rows,
        cvImg.step1(),
        System::Drawing::Imaging::PixelFormat::Format24bppRgb,
        (System::IntPtr)cvImg.data);
    }
```

```
    //灰度
    else if(cvImg.channels()==1)
    {
        cv::Size size = cvImg.size();
        bmpImg = gcnewBitmap(size.width,size.height,cvImg.step,
             System::Drawing::Imaging::PixelFormat::Format8bppIndexed,(System::
             IntPtr)cvImg.data);
    }

    return(bmpImg);
}
```

在深度学习应用使用C#平台时,这两个转换是花费时间较多。我们可以考虑采用拷贝数据地址的方式,这样可以大大提高转换效率,缩短转换时间。

14.3 字符串转换

在C#中的字符串类型System::String与C++的字符串类型string的转换代码如下所示。

```
System::String ConvertToSystemString(string str)
{
    return(gcnew System::String(str.c_str()));
}


string ConvertToCppstring(System::String str)
{
    char * chars2 = (char * )(Marshal::StringToHGlobalAnsi(str)).ToPointer();
    string s = chars2;
    Marshal::FreeHGlobal(IntPtr((void * )chars2));
    return s;
}
```

14.4 框架设计

整体框架主要分为三部分:首先是C#部分,这是程序的入口,完成图像的采集和载入。其次是CLR部分,将图像和字符串的格式进行转换。最后是C++的算法部分,这是基于C++和OpenCV的CV::Mat格式的算法执行。如图14-2所示。

有了该框架,C#每次就可以采用DLL装载的方式,将CLR转换过来的格式让C#和C++互联。CLR中间连接器可以让界面开发者采用C#,算法开发者采用C++,这样就可以将界面开发与算法开发完全分离,大大降低了开发成本,同时也提高了程序运行的效率。

图 14-2 C#、CLR、C++平台框架

思 考 题

简答题

1. 为什么自动化设备的上位机都会选择 C# 语言来开发？
2. C# 平台下，如何将 Python 和 C++ 进行深度融合？
3. 为什么需要将 C# 的图像格式与 C++ 格式进行转换？

第 15 章

基于 C#深度学习焊点缺陷检测

15.1 C#深度学习开发流程

深度学习与传统的机器视觉的融合,也就是图像的前处理与后处理的融合。本方法是将深度学习的方法,将图像处理的前处理用深度学习的方法代替,这样在开发图像处理软件时,能提高 80% 以上的开发速度,并且降低开发者 80% 以上的成本。因为深度学习可以代替图像处理 80% 的任务(图像前处理部分)的同时,将图像开发的风险降为 0,这样就可以在开发成本降低为 80% 的同时,对于图像处理的人员要求降低 80%。整个开发流程如图 15-1 所示。

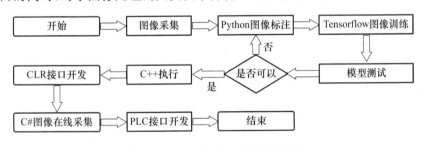

图 15-1 C#深度学习开发平台框架

在非标定制自动化的设计当中,大多数企业会将 C#作为图像的采集和 GUI 界面的开发,Python 作为图像标注和深度学习的开发语言,C++作为高效执行的语言。

图像定位、测量、检测和识别的核心算法就是区域的提取和特征提取,本方法通过深度学习的网络,将图像区域的提取和特征提取采用统一的深度学习的模型。具体的过程描述为:通过 labelme 进行标注后,进行基于 Unet 网络模型的训练,从而完成的区域定位和特征的提取。

15.2 焊点检测任务

焊点检测的需求是检测焊点的位置、个数、面积大小和焊点的缺陷四个方面。总体方法是采用人工标注焊点,通过深度学习网络训练后,获得焊点区域模型。这样就可以确定焊点的位置、个数和面积大小,然后根据每个焊点的位置,获得整体焊点的区域后,采用人工标注的方法,确定缺陷的位置进行标注和训练,以此获得缺陷模型。获得的一张原始图片如图 15-2 所示。

图 15-2 通过海康彩色相机 1200 万像素采集焊点图像

标注后，采用 json 解析后的目标图像如图 15-3 所示。

图 15-3 人工标注后解析获得焊点图像

通过深度学习网络训练后，模型测试的效果如图 15-4 所示。

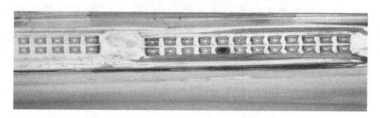

图 15-4 采集有焊点缺陷的焊点图像

通过深度学习获得的缺陷和基本几何信息的结果图像（包括焊点大小、个数、面积 OK、缺陷检测 NG 的结果）如图 15-5 所示。

图 15-5 通过深度学习检测缺陷 NG 的图像

一张原始图像的焊点大小、个数、面积 OK，没有缺陷 OK 原始图像，如图 15-6 所示。

图 15-6 采集没有焊点缺陷的焊点图像

一张测试结果图像(包括焊点大小、个数、面积)均 OK,没有缺陷 OK 的总的结果图像如图 15-7 所示。

图 15-7 检测没有焊点缺陷的焊点图像

采用传统的图像处理方式进行图像的定位和特征提取,是图像前处理的核心。由于每一张图像的光照条件和成像的效果都不一样,导致在图像处理的过程中,需要对不同的图像采用不同的预处理的方法。这样,在开发图像的应用方面,就存在开发方法多样性、开发周期长、维护成本高的问题。

15.3 深度学习焊点检测分析

图 15-8 所示的上半部分的图像是焊点大小、个数、面积均 OK,下半部分的黑白图像是深度学习推理出来的焊点缺陷图片,尽管焊点不是特别完美,但是也在产品质量的接收范围内,所以焊点判断为没有焊点缺陷,所以总体判断为 OK 图片。

图 15-8 检测没有焊点缺陷的焊点图像

图 15-9 所示的上半部分的图像是焊点大小、个数、面积均 OK,下半部分的黑白图像是深度学习推理出来的焊点缺陷图片,可以看出有明显的焊点缺陷,严重地影响了产品的质量,所以焊点判断为有焊点缺陷,所以总体判断为 NG 图片。

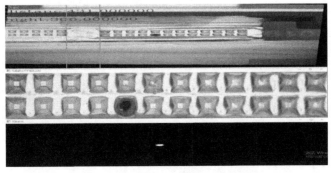

图 15-9 检测有焊点缺陷的焊点图像

对于产品的缺陷检测,误判率、漏判率和执行时间是评价一个设备的核心指标。为了达到这三个指标,产品在试产阶段通常要经过四个阶段:检测标准、样本收集、模型训练、在线部署。这四个阶段正常情况下需要花费四周的时间,然后可以进行产品上线测试。

思 考 题

简答题

1. C#语言来开发深度学习项目的流程是什么?
2. 视觉检测项目的三个重要的指标是什么?
3. 深度学习项目开发的四个步骤是什么?

第 16 章
基于 Pyqt5 和 OpenCV 图像处理平台

16.1 说在前面

学习者只需要有 Python 的编程基础就可以完成项目的所有任务。整个系统是由 Pyqt5 作为开发界面，Python 版本的 OpenCV 作为图形开发的工具，是一个机器视觉图像采集与算法开发的平台。

16.2 编写目的

机器视觉与深度学习的项目的开展，基于 Python 语言的开发框架 Pyqt5 的开发模式，既可以满足机器视觉采集图像的要求，又可以满足算法的开发的要求。

16.3 产品范围

机器视觉的两个关键因素：采集图像与算法设计。本产品是完成采集图像的任务，包括相机的选型、镜头的配置、光源的选择，以及整体的打光方案的设计与测试能力的评估。

16.4 阅读建议

建议先阅读完软件开发平台要求，配置好运行环境，并保证能正常打开软件，然后了解下图像处理的功能，最后作为学习者，可以去完善的功能。

16.5 运行环境、库、GPU 要求

操作系统为 Windows 10 版本。编程环境为 IDE，采用 VScode，开发工具和驱动工具为 Python 和 OpenCV。配置参考网络链接为：https://blog.csdn.net/weixin_40283570/article/details/97111691。

16.6 环境配置

(1) Python 离线安装方法，下载 Python 安装文件后，进行安装。
(2) Pip 版本升级：python-m pip install--upgrade pip，如图 16-1 所示。

图 16-1 升级 pip 版本为当前版本

(3) OpenCV 在线安装 pip install opencv-python，如图 16-2 所示。

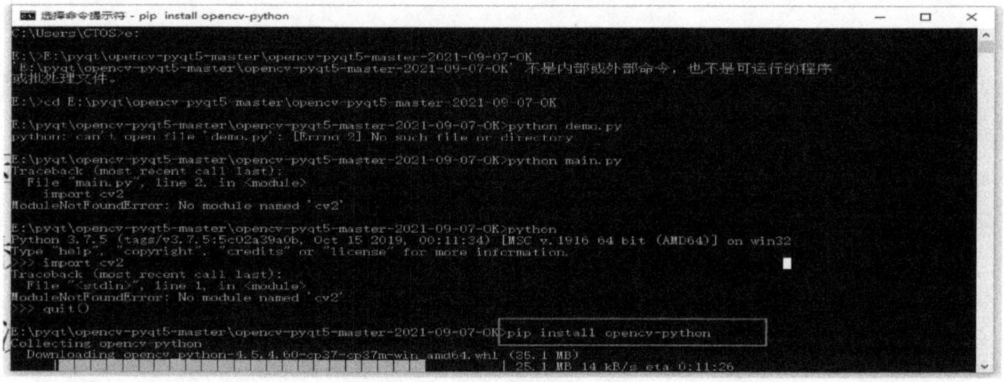

图 16-2 安装 Python 版本的 OpenCV

(4) 在线安装 Pyqt5 方法：pip install PyQt5，如图 16-3 所示。

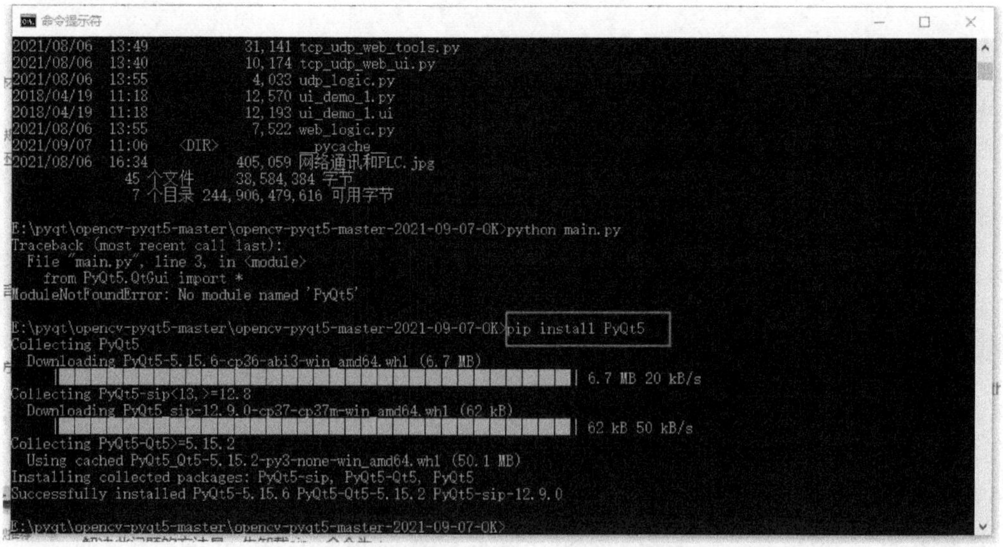

图 16-3 安装 Python 版本的 Qt

第 16 章 基于 Pyqt5 和 OpenCV 图像处理平台

(5) 安装图像化界面：pip install matplotlib，如图 16-4 所示。

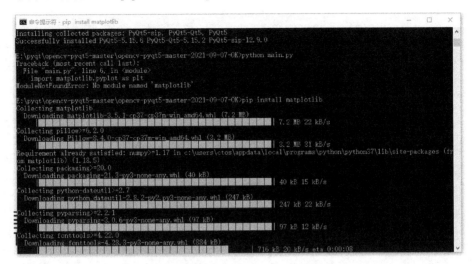

图 16-4 安装图像化界面库 matplotlib

(6) 安装相机的驱动：pip install pypylon，如图 16-5 所示。

图 16-5 安装 Python 版本的 Blaser 相机

(7) 显示界面 python main.py，如图 16-6 所示。

图 16-6 Python 视觉平台启动界面

（8）在集成编程环境 Visual Studio Code 进行编辑和运行，如图 16-7 所示。

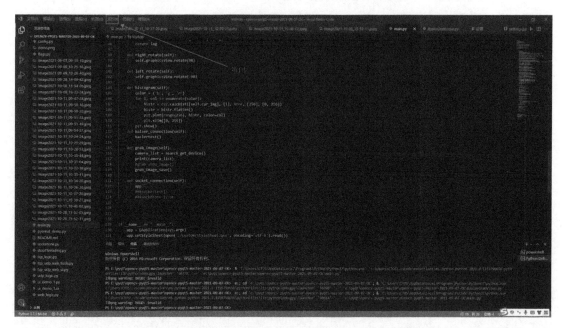

图 16-7　Python 源码采用 Visual Studio Code 进行编辑

（9）选择项为 Python 即可，如图 16-8 所示。

图 16-8　选择 Python 作为运行的环境

（10）选择项为 Python 然后就出现程序运行结果的画面，如图 16-9 所示。

第16章 基于 Pyqt5 和 OpenCV 图像处理平台

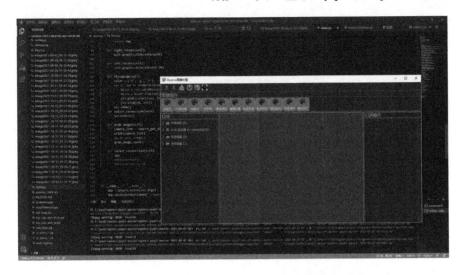

图 16-9　源码启动 Python 开发平台界面

16.7　框架介绍

整个程序的入口是 main.py，其中主要的功能模块分为三部分：通信功能、实时图像显示功能、图像采集功能和图像算法分析功能。如图 16-10 所示。

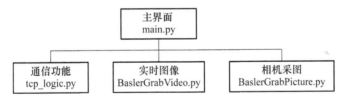

图 16-10　Python 框架功能图

16.8　未来计划

（1）目前软件的相机取图只有一个相机采集图像的功能，后面需要进行双相机的图像功能的开发。

（2）目前软件只是用 OpenCV 的算法完成图像处理的功能，后面将深度学习的功能添加进去。

思　考　题

简答题

1. 机器视觉与图像处理的关系是什么？
2. 图像处理的基本步骤有哪些？

第 17 章
基于 C++ 视觉软件系统构建

本章介绍基于 C++ 视觉软件系统，主要包括：日志跟踪、图像采集、光源控制。本章就视觉软件的三个方面进行详细的软件说明。

17.1 基于 QThread 的日志跟踪

在日志处理的头文件 qlog.h 中，QLog 是继承 QThread，最核心的代码是将 run 函数进行重写。为了避免多线程日志写入的时候的一致性，采用互斥 QMutex 和信号量 QSemaphore 的方式。代码如下：

```
#ifndef QLOG_H
#define QLOG_H
#include<stdio.h>
#include<stdlib.h>
#include<QDebug>
#include<QMessageBox>
#include<QFile>
#include<QFileInfo>
#include<QThread>
#include<QList>
#include<QSemaphore>
#include<QMutex>

class QLog: public QThread
{
    Q_OBJECT
public:
    QLog();
    void write(const QString &msg);
    virtual void run();
private:
```

```
    QMutex m_mutex;
    QList<QString> m_msg;
    QSemaphore m_synSem;
};

#endif // QLOG_H
```

在日志处理的实现 qlog.cpp 中,在写入日志的函数 write 中,采用加锁和解锁的方式来完成日志的写入。在重写 run 函数中,采用信号量的方式,让多个线程调用日志写入的时候,保证只有一个写入正在执行。代码如下。

```
#include "qlog.h"
#include<QMutex>
#include<QFile>
#include<QApplication>
#include<QDate>
#include<QDebug>

QLog::QLog(): m_synSem(0)
{
}

void QLog::write(const QString &msg)
{
    m_mutex.lock();

    QString current_date_time = QDateTime::currentDateTime().toString("yyyy-MM-dd hh:mm:ss ddd");
    QString current_date = QString("Jesus God! (%1)").arg(current_date_time);
    QString message = QString(" %1 %2").arg(msg).arg(current_date);

    m_msg.push_back(message);

    m_mutex.unlock();
    m_synSem.release();
}

void QLog::run()
{
    qDebug()<<"run()";
    while(true)
    {
        m_synSem.acquire();

        m_mutex.lock();
```

```
            if(m_msg.isEmpty())
            {
                continue;
            }

            QString message = m_msg.front();
            m_msg.pop_front();

            QFile file("log.txt");
            file.open(QIODevice::WriteOnly | QIODevice::Append);
            QTextStream text_stream(&file);
            text_stream << message <<"\r\n";
            file.flush();
            file.close();
            m_mutex.unlock();
        }
}
```

在日志文件的调用方法中，首先采用 start 启动线程，然后调用 write 函数，将需要写入的日志进行写入。代码如下。

```
void MainWindow::on_pushButton_clicked()
{
    m_log->start();
    m_log->write("CTK EXE start...");

    ui->textBrowser->insertPlainText("CTK EXE start...\n");

}
```

17.2 基于海康工业相机的图像采集

图像采集是机器视觉软件中最为核心的部分。海康工业相机是中国自主品牌中最具影响力的工业相机。海康工业相机的 API 如图 17-1 所示。

海康工业相机的类的构建的方法如图 17-2 所示。

海康相机的调用方法分为三步，第一步创建一个类指针，第二步输入相机的编号，第三步析构函数处理退出和停止。

图 17-1　海康工业相机的 API

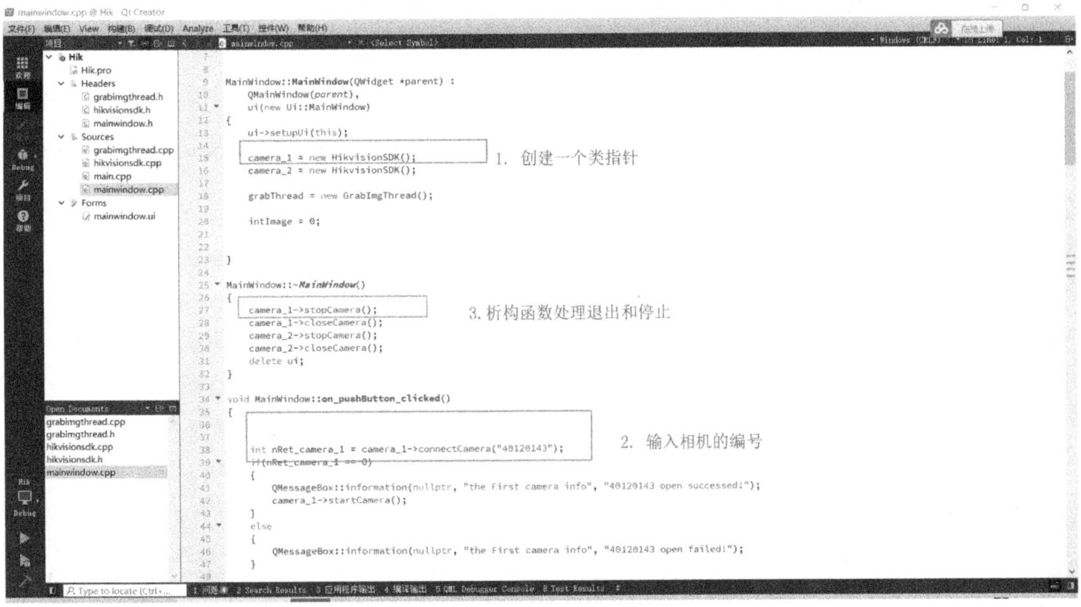

图 17-2　海康相机的调用方法

17.3　基于 Basler 工业相机的图像采集

Basler 是一个德国的工业相机品牌，其采集图像可以分为以下五个步骤。
（1）自动获取相机。
（2）打印相机的名称。

(3) 获得相机的图像。
(4) 获取图像。
(5) 图像转换。

德国 Basler 工业相机的调用方法分别如图 17-3 和图 17-4 所示。

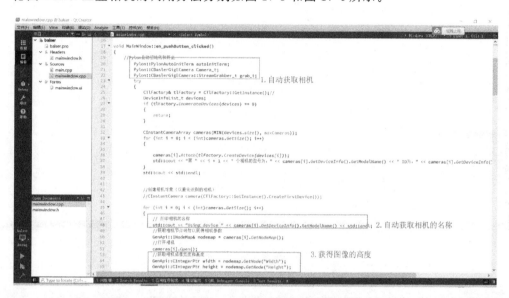

图 17-3　德国 Blaser 工业相机的调用方法的第 1、2、3 步

图 17-4　德国 Blaser 工业相机的调用方法的第 4、5 步

在选择工业相机时，推荐使用海康的工业相机，除非是客户指定，否则使用德国的 Blaser 相机。同时，海康相机的网络相机和 Blaser 的网络相机是通用的，也就是使用海康的驱动，也能使用德国 Blaser 的相机。

17.4 基于串口通信和光源控制系统

串口通信是一个很常见的通信方式,有很多厂家的光源是通过串口通信来完成。采用 Qt 的窗口通信的类 QtSerialPort,这样在调用的时候,就能够跨平台的调用。为了实时显示串口通信的数据,我们采用定时器的类 QTimer 来完成。代码如下。

```
#ifndef SERIALPORTDIALOG_H
#define SERIALPORTDIALOG_H

#include<QDialog>
#include<QtSerialPort/QSerialPort>
#include<QtSerialPort/QSerialPortInfo>
#include<QThread>
#include "my_qserial.h"
namespace Ui {
class serialportDialog;
}
class serialportDialog: public QDialog
{
    Q_OBJECT
public:
    explicit serialportDialog(QWidget * parent = nullptr);
    ~serialportDialog();
    void on_btnOpen_clicked();
    void on_btnSend_clicked();
public slots:
    void readComDataSlot();
    void on_btn_Connect_clicked();
private:
    Ui::serialportDialog * ui;
    my_qserial * local_serial;
    void init();
    QSerialPort * my_serialport;
    QTimer * timer;
};

#endif // SERIALPORTDIALOG_H
```

在串口通信的 CPP 实现中,将串口通信的波特率、奇偶校验、是否设置停止位和光源控制器的串口通信协议保持一致。代码如下。

```cpp
oid serialportDialog::init()
{
    qDebug()<<"thread_sig thread"<< QThread::currentThreadId();
    foreach(const QSerialPortInfo & info, QSerialPortInfo::availablePorts())
    {
        qDebug()<<"Name:"<< info.portName();
        qDebug()<<"Description:"<< info.description();
        qDebug()<<"Manufacturer:"<< info.manufacturer();


        QSerialPort serial;
        serial.setPort(info);
        if(serial.open(QIODevice::ReadWrite))
        {
            ui->cmbPortName->addItem(info.portName());
            serial.close();
        }
    }

    QStringList baudList;
    QStringList parityList;
    QStringList dataBitsList;
    QStringList stopBitsList;
    QStringList streamControlList;

    baudList<<"50"<<"75"<<"100"<<"134"<<"150"<<"200"<<"300"<<"600"<<"1200"<<"1800"<<
    "2400"<<"4800"<<"9600"<<"14400"<<"19200"<<"38400"<<"56000"<<"57600"<<"76800"<<
    "115200"<<"128000"<<"256000";

    ui->cmbBaudRate->addItems(baudList);
    ui->cmbBaudRate->setCurrentIndex(12);

    parityList << u8 << u8 << u8;
    parityList << u8;
    parityList << u8;

    ui->cmbParity->addItems(parityList);
    ui->cmbParity->setCurrentIndex(0);

    dataBitsList<<"5"<<"6"<<"7"<<"8";
    ui->cmbDataBits->addItems(dataBitsList);
    ui->cmbDataBits->setCurrentIndex(3);
```

```
    stopBitsList<<"1";
    stopBitsList<<"1.5";
    stopBitsList<<"2";

    ui->cmbStopBits->addItems(stopBitsList);
    ui->cmbStopBits->setCurrentIndex(0);

    streamControlList<<u8;
    streamControlList<<u8;

    ui->cmbStreamControl->addItems(streamControlList);
    ui->cmbStreamControl->setCurrentIndex(0);

    ui->readButton->setCheckable(true);

}
```

C++机器视觉软件平台是否考虑到深度学习部分,取决于GPU类型。如果GPU是1050ti、2080ti等系列,平台统一采用如下的软件配置:Win 10 64位、VS 2019、QT 5.15.0、Cmake 3.22.1、OpenCV 4.5.3、cuda 10.1、Tensorflow 2.3.0。

如果GPU是3050ti、3090等系列,平台统一采用如下的软件配置:Win 10 64位、VS 2019、QT 5.15.0、Cmake 3.22.1、OpenCV 4.5.3、cuda 11.2、Tensorflow 2.6.0。

除了上述内容外,机器视觉软件平台与PLC的通信部分和数据可视化、ERP系统的对接,都是视觉软件非常重要的部分。

思 考 题

简答题

1. 互斥和信号量的关系是什么?
2. 图像采集中的软触发和硬触发的区别是什么?

第 18 章
深度学习用于焊缝检测实验报告

18.1 检测条件

焊缝检测与测量项目,相机曝光由高低曝光两组图片组成,像素点的大小为 0.03 mm,焊缝检测的宽度小于 0.2 mm。电脑配置:CPU 是 i7-7700,内存是 16G,GPU 是 1050TI,操作系统是 Win 10。

18.2 测试结果

测试条件:测试由 500 张图片作为总体样本。OK 样本为 453 个,NG 样本为 47 个。
测试结果:453 个 OK 样本的测试结果为 447 个 OK,6 个 NG。
47 个 NG 样本的测试结果为 44 个 NG,3 个 OK。
本次测试按照总样本 500 个来计算,过检率 6/500=1.2%,漏检率 3/500=0.6%。

18.3 存在问题

问题一:打光的方式采用低曝光与高曝光的组合,导致深度学习在判断时,基准图像会出现两个基准。
问题二:对于没有出现过的焊缝样本量数量太少,本次 NG 漏检的图片只有一个样本,导致深度学习在样本学习的时候,缺陷样本太少。
原始图片 OK,测试结果 OK 的图片,被统称为判断正确图片,如图 18-1～图 18-6 所示。
原始图片 OK,测试结果 NG 的图片,被统称为过检图片,如图 18-7～图 18-12 所示。
原始图片 NG,测试结果 NG 的图片,被统称为判断正确图片,如图 18-13～图 18-27 所示。
原始图片 NG,测试结果 OK 的图片,被统称为漏检图片,如图 18-28、图 18-29 和图 18-30 所示。

第18章 深度学习用于焊缝检测实验报告

  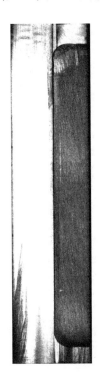 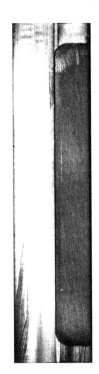

图 18-1　原始图片(左)和测试结果(右)　　　　图 18-2　原始图片(左)和测试结果(右)

 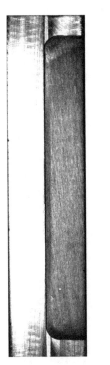 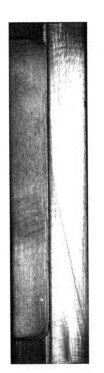 

图 18-3　原始图片(左)和测试结果(右)　　　　图 18-4　原始图片(左)和测试结果(右)

图 18-5 原始图片(左)和测试结果(右)　　　　图 18-6 原始图片(左)和测试结果(右)

  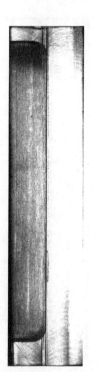 

图 18-7 原始图片(左)和测试结果(右)　　　　图 18-8 原始图片(左)和测试结果(右)

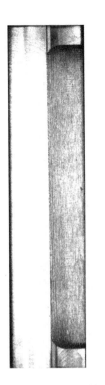  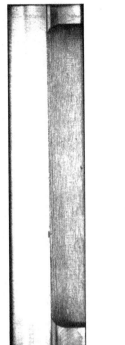 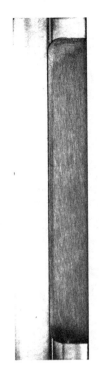 

图 18-9　原始图片(左)和测试结果(右)　　　　图 18-10　原始图片(左)和测试结果(右)

  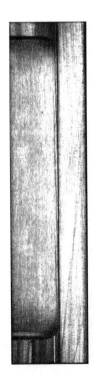 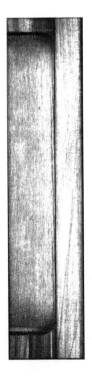

图 18-11　原始图片(左)和测试结果(右)　　　　图 18-12　原始图片(左)和测试结果(右)

 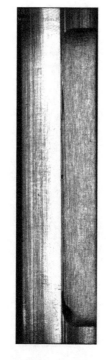  

图 18-13 原始图片(左)和测试结果(右)　　　　图 18-14 原始图片(左)和测试结果(右)

  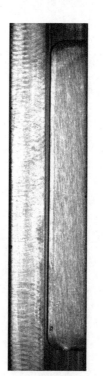 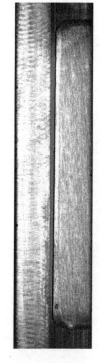

图 18-15 原始图片(左)和测试结果(右)　　　　图 18-16 原始图片(左)和测试结果(右)

 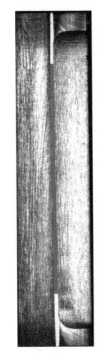 

图 18-17　原始图片(左)和测试结果(右)　　　　图 18-18　原始图片(左)和测试结果(右)

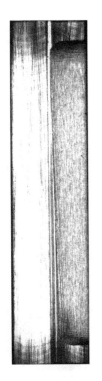   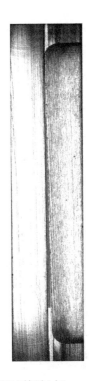

图 18-19　原始图片(左)和测试结果(右)　　　　图 18-20　原始图片(左)和测试结果(右)

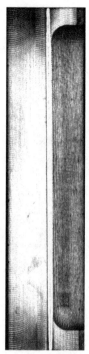 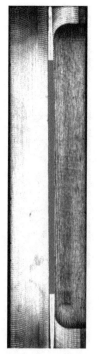 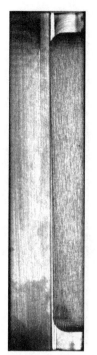 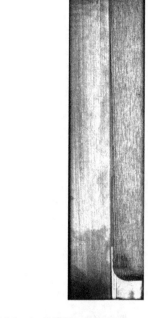

图 18-21　原始图片(左)和测试结果(右)　　　　图 18-22　原始图片(左)和测试结果(右)

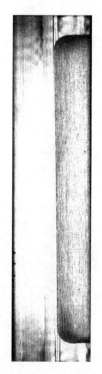 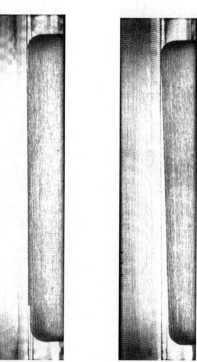 

图 18-23　原始图片(左)和测试结果(右)　　　　图 18-24　原始图片(左)和测试结果(右)

 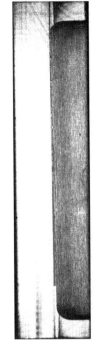 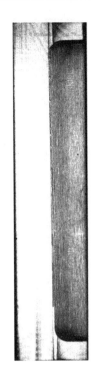

图 18-25 原始图片(左)和测试结果(右)　　　　图 18-26 原始图片(左)和测试结果(右)

  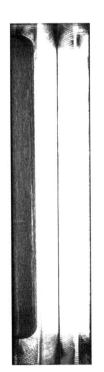 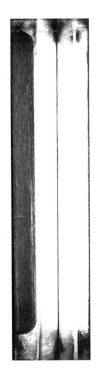

图 18-27 原始图片(左)和测试结果(右)　　　　图 18-28 原始图片(左)和测试结果(右)

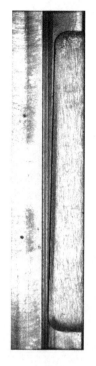 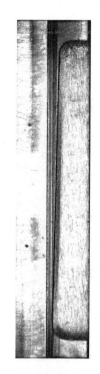
图 18-29 原始图片(左)和测试结果(右)

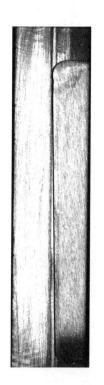 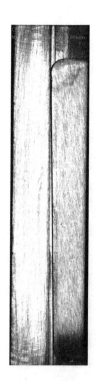
图 18-30 原始图片(左)和测试结果(右)

18.4 改善方法

改善方法 1：图像采集质量的提升。

改善方法 2：采用中等曝光的采集图像的方式，将图像的学习基准从两种曝光调整为一种曝光模式。

改善方法 3：在现场收集更多的缺陷样本后，进行模型优化。

18.5 实验结论

采用深度学习方法可以完成焊缝检测，测量的精度约为两个像素，重复精度需要以实际数据为准。

思 考 题

简答题

1. 深度学习检测实验报告的模板是什么？
2. 如何从图像优化、样本量和算法优化三个方面详细描述深度学习的改善措施？

第 19 章
Win 10 环境下深度学习训练环境构建

本章介绍在 Win 10 环境下，深度学习模型训练是如何搭建的。19.1 节介绍了 Python 编程语言的安装，19.2 节讲解了 labelme 标注软件的安装，19.3 节重点讨论了科学计算中 Numpy 模块的安装，19.4 节介绍了开发环境 Microsoft Visual Studio 2019 系统的安装，目的是采用 C++ 高效解析标注后的 json 文件，用来生成原始图像和目标图像。19.5 节介绍了 Microsoft Visual Studio Code 系统安装和 Python 语言的集成开发环境。19.6 节介绍了 Tensorflow 的 GPU 版本的安装，用来进行深度学习模型的训练与测试。19.7 节介绍了 OpenCV 的安装，以作为图像处理的开源库。19.8 节介绍了 GPU 的 C++ 的开发库 CUDA，当进行深度学习推理时，提供 C++ 的 GPU 的接口驱动。19.9 节介绍了 Cudnn 深度学习算法库，将对应的 include 和 lib 目录拷贝到 CUDA 目录下即可。19.10 节介绍了 Keras 在 Tensorflow 2.0 版本以下是需要独立安装的，Keras 是 Tensorflow 更为上层的抽象和包装，更容易开发应用系统。

19.1 Python

本项目所需的 Python 版本为 3.7.5，在 Python 官网进行下载，如图 19-1 所示。找到 Win 10 对应的版本，下载完成后双击打开安装包进行以下步骤，如图 19-2、图 19-3、图 19-4 和图 19-5 所示。

图 19-1　下载 Python 3.7.5 版本

C++语言与机器视觉编程实战

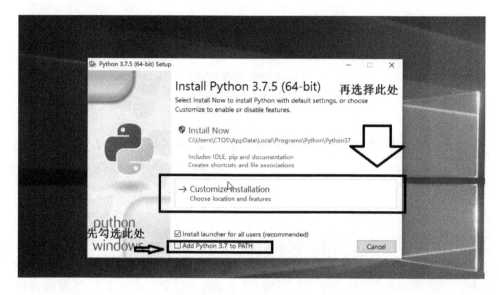

图 19-2　强烈建议勾选"Add Python 3.7 to PATH"

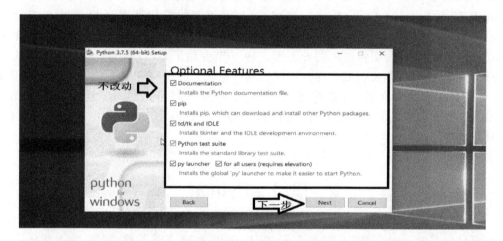

图 19-3　单击"Next"

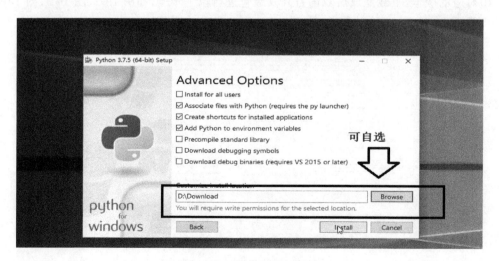

图 19-4　选择下载的路径

|第 19 章| Win 10 环境下深度学习训练环境构建

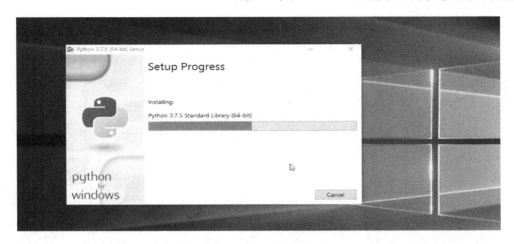

图 19-5　Python 的安装过程

完成以上步骤后进行下一步验证，打开命令提示符，如图 19-6 所示。

图 19-6　从 Win 10 进入命令行

输入"python"，如图 19-7、图 19-8 和图 19-9 所示。

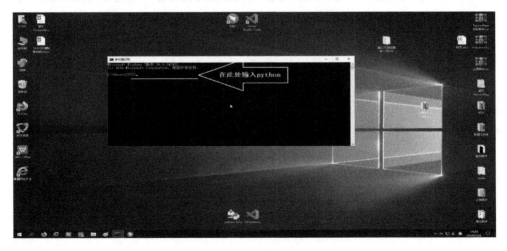

图 19-7　cmd 命令进入命令行

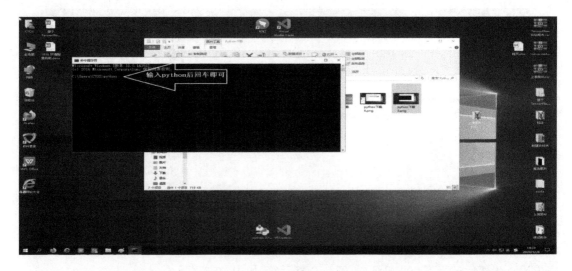

图 19-8　从命令行输入"python"

图 19-9　查看 Python 版本

按以上步骤来验证 Python 是否安装完成。

19.2　labelme

安装 labelme 时需要用到 Windows 命令行以成功安装 Python。打开命令行，如图 19-10 所示。

在命令行输入"pip install labelme"，在线安装 labelme，如图 19-11 所示。

labelme 安装完毕，在命令行输入"labelme"并回车，如图 19-12 所示。

第 19 章 Win 10 环境下深度学习训练环境构建

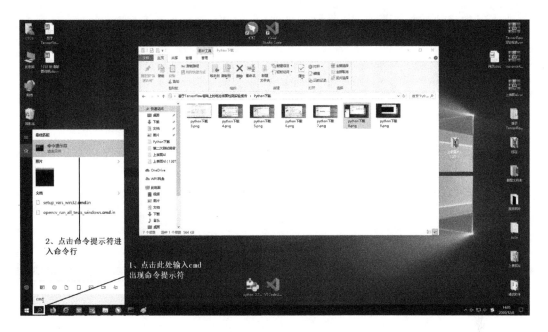

图 19-10　进入命令行

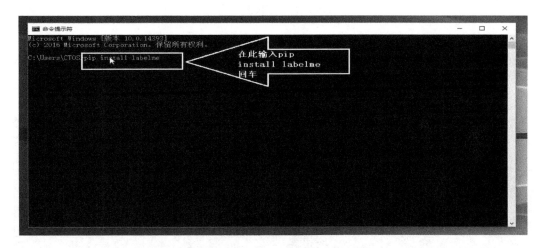

图 19-11　在线安装 labelme

图 19-12　从命令行输入"labelme"

.labelme 的启动界面如图 19-13 所示。

图 19-13　labelme 的启动界面

19.3　Numpy

安装 Numpy 时需要先成功安装 Python，打开命令行，输入"pip install numpy"即可，如图 19-14 所示。

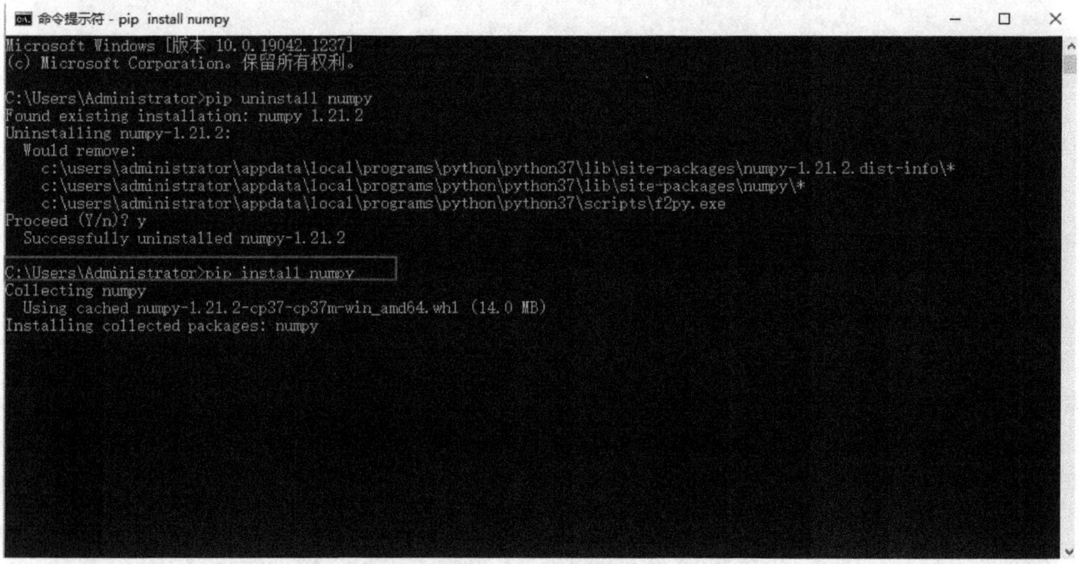

图 19-14　成功安装 Numpy

第 19 章 | Win 10 环境下深度学习训练环境构建

当出现如图 19-15 所示的界面时则表示安装成功。

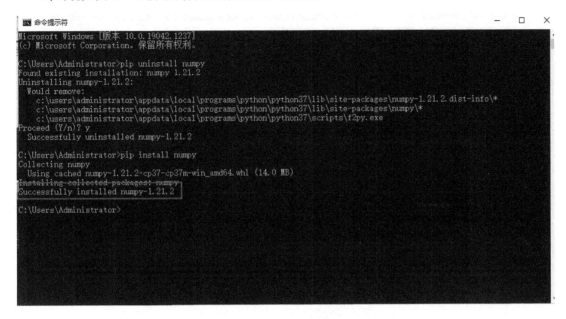

图 19-15 成功安装 Numpy

19.4 Microsoft Visual Studio 2019

下载完成后打开安装包,如图 19-16 所示。

图 19-16 下载完成后单击"继续",安装 Microsoft Visual Studio 2019

等待安装,如图 19-17 所示。

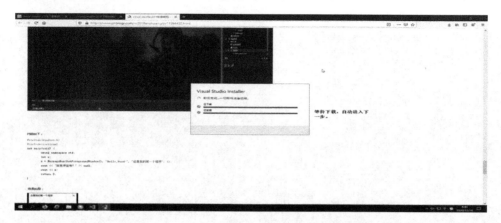

图 19-17　Microsoft Visual Studio 2019 的安装过程

勾选打勾的四个选项,如图 19-18 所示。

图 19-18　勾选打勾的四个选项

等待其自行下载完毕后即可使用,如图 19-19 所示。

图 19-19　在线式下载安装 Microsoft Visual Studio 2019

19.5　Microsoft Visual Studio Code

选择同意安装 Microsoft Visual Studio Code，并打开安装包，如图 19-20 所示。

图 19-20　选择同意安装 Microsoft Visual Studio Code

Microsoft Visual Studio Code 安装完毕，如图 19-21 所示。

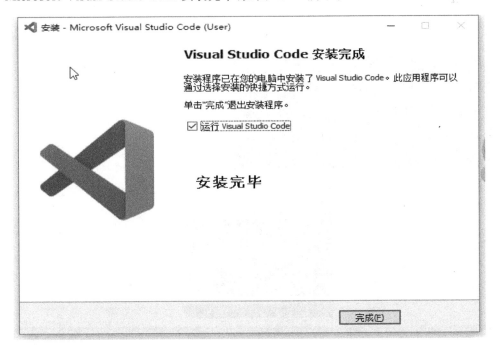

图 19-21　Microsoft Visual Studio Code 安装完毕

19.6 TensorFlow

版本需求为 TensorFlow-GPU==1.15.2，打开命令行输入"pip install TensorFlow-GPU==1.15.2"，等待下载，如图 19-22 左上角所示。

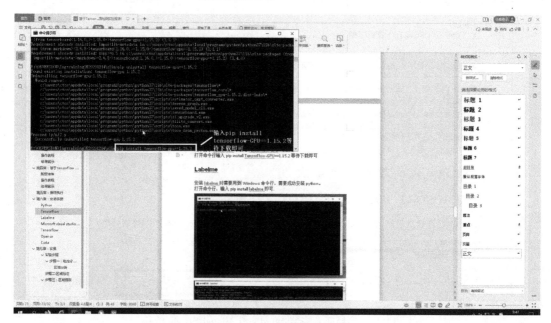

图 19-22 等待下载

下载完成后的提示如图 19-23 所示。

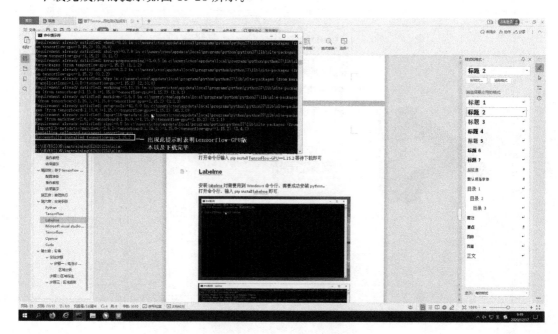

图 19-23 下载完成后的提示

19.7 OpenCV

所需版本为 OpenCV-4.3.0，打开下载完成的安装包，如图 19-24 所示。

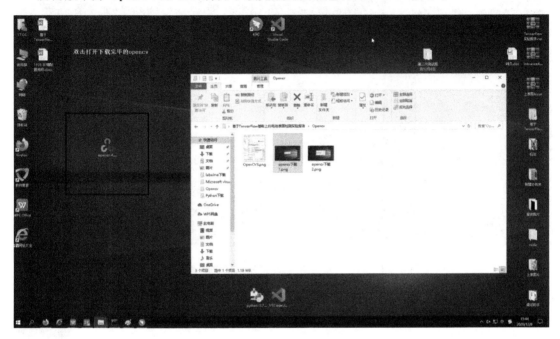

图 19-24　OpenCV 下载完毕

选择存放路径，单击"Next"，如图 19-25 所示。

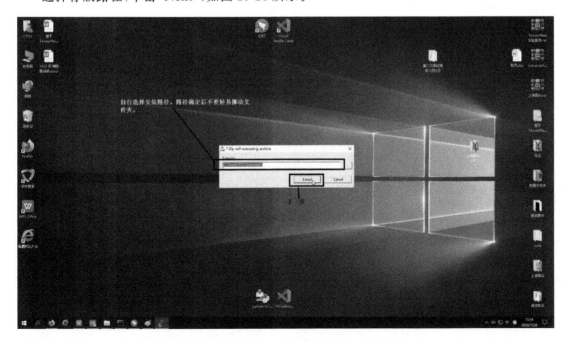

图 19-25　选择存放路径

OpenCV 自动解压到指定的路径,如图 19-26 所示。

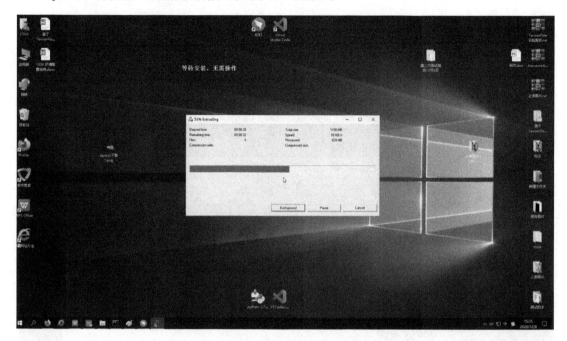

图 19-26　OpenCV 自动解压到指定的路径

在环境变量中,添加 OpenCV 的路径到 Path 中,如图 19-27 所示。

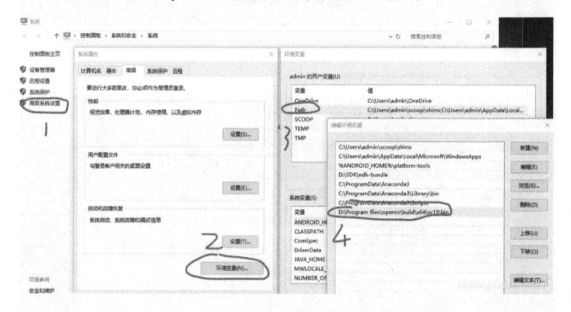

图 19-27　OpenCV 环境变量配置

19.8　Cuda

所需版本为 Cuda 10.1 和 Cuda 10.0,在官网自行下载 Cuda 10.1、Cuda 10.0 安装包。

第 19 章 | Win 10 环境下深度学习训练环境构建

在官网选择下载 Cuda 页面如图 19-28 所示。

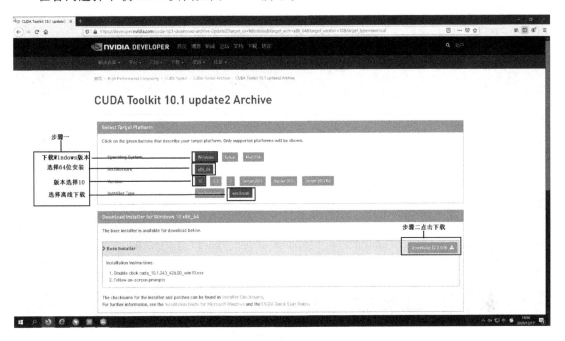

图 19-28　选择下载 Cuda

等待下载完毕后，打开安装包，选择 Cuda 安装路径，如图 19-29 所示。

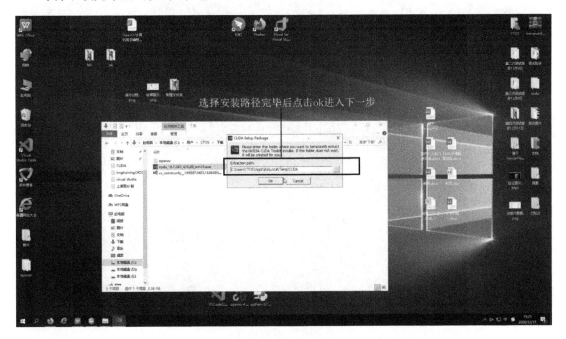

图 19-29　选择 Cuda 安装路径

等待进度达到 100%,如图 19-30 所示。

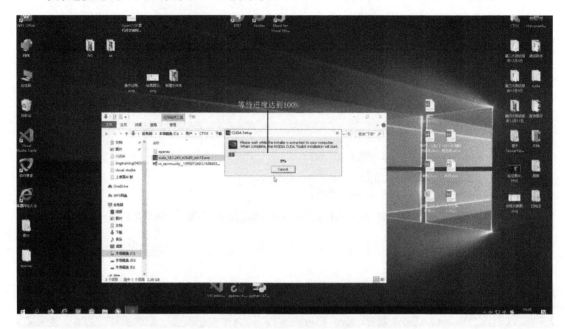

图 19-30　Cuda 安装进度

等待系统环境检测,如图 19-31 所示。

图 19-31　系统检测

选择自定义安装后,操作如图 19-32 所示。

图 19-32　取消此处勾选

安装完成后,打开命令行验证 CUDA 安装是否成功,在命令行输入 nvcc-version,如图 19-33 所示。

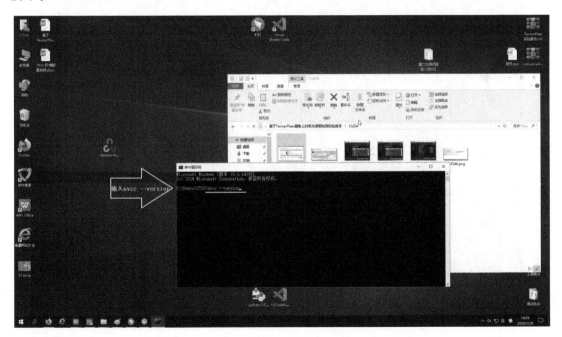

图 19-33　验证 Cuda 是否安装成功

在命令行输入"nvcc--version"后出现图 19-34 框中文字则进入下一步，手动输入"set cuda"并回车。

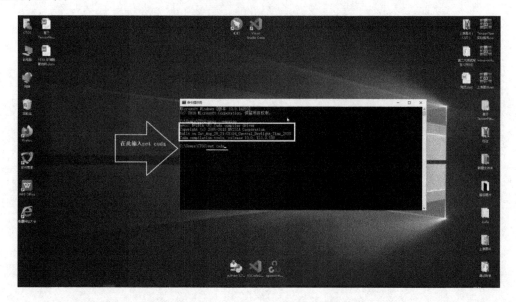

图 19-34　在命令行输入"set cuda"

在命令行输入"set cuda"后，出现的结果与下图一致则说明 CUDA 安装完毕，如图 19-35 所示。

图 19-35　Cuda 安装完毕

19.9　Cudnn

所需版本为 Cudnn=10.1 和 Cudnn=10.0，首先在官网自行下载 CUDA 10.1 对应版本的

Cudnn 10.1 与 Cuda 10.0 对应版本的 Cudnn 10.0，然后进行下载解压。将下载的 Cudnn 压缩包解压后，将 Cuda 文件夹中的 bin、include、lib 三个文件夹进行替换即可，如图 19-36 所示。

图 19-36　Cudnn 拷贝 bin、include、lib 到 Cuda

19.10　Keras

所需版本为 Keras==2.3.0，打开命令行，如图 19-37 所示。

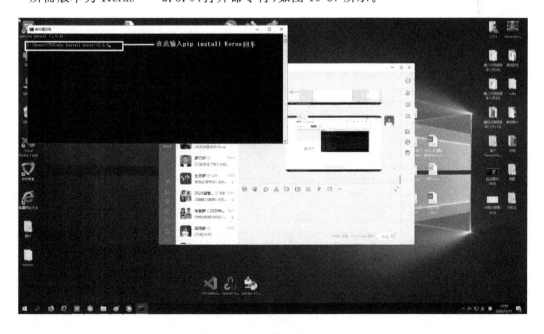

图 19-37　命令行下安装 Keras

输入"pip install Keras==2.3.0"后按回车键等待下载即可,如图 19-38 所示。

图 19-38　选择同意安装 Visual Studio Code

至此,基于 Win 10 深度学习开发平台的整个系统安装过程全部完成。

思 考 题

简答题

1. Cuda 的版本是如何选择的?
2. Cuda 和 Cudnn 是什么关系?
3. 如何配置 OpenCV 的环境变量?

第 20 章 视 觉 方 案

视觉方案是根据客户的需求(检测的要求:精度、指标、标准),拿到客户的样品后,进行打光测试,形成的视觉评估报告。这就是对于光源、相机、镜头的硬件选型和安装方法,供装备人员安装和调试时使用。

20.1 项目需求

举一个电池行业中的级片检测的例子,对前极组上下两个大面进行检测。客户样品尺寸示意图如图 20-1 所示。在产品检测中,需要兼容的范围见表 20-1。

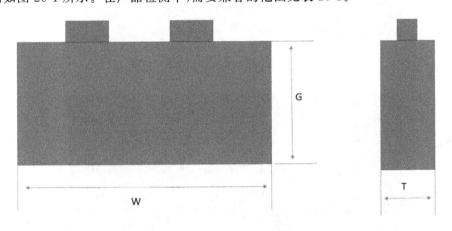

图 20-1 客户样品尺寸示意图

表 20-1 宽度和长度范围

尺寸	兼容范围/mm
W	117~247
G	80~130

检测要求:利用色差检测≥0.1 mm 以上的金属异物、黑点和破损。硬件方案:使用 1 200 W (4 096×3 000 分辨率的相机)全局彩色相机,镜头视野 260 mm×160 mm,像素精度为 260/4 000=0.02 mm/pixel。其中,分区光源的内直径为 280 mm,加装直径 280 mm 碗状光源。

20.2 打光图片

图 20-2~图 20-6 所示为分区光源五个方向光源的图像采集:上区域、下区域、左区域、右区域、全区域。

图 20-2 分区光源打光图片(上区域光源亮)

图 20-3 分区光源打光图片(下区域光源亮)

图 20-4 分区光源打光图片(左区域光源亮)

图 20-5 分区光源打光图片(右区域光源亮)

图 20-6 分区光源打光图片(所有区域光源亮)

20.3 结论分析

以 0.1 mm 精度需求,设计 260 mm×160 mm 的视野范围,配上 4 096×3 000 分辨率的相机,可实现 0.063 像素精度,满足公差值 30%的设计要求。由于上、下面都需要检测,最大电芯单面需要拍照一次,需要设计单轴移动一次,进行另一个单面拍照一次。

其中的风险点如下。

(1) 由于外观检测算法复杂,要求严格,单次检测时间需要 0.8 s。
(2) 产品需要放置在固定平台进行外观检测。
(3) 需要设计单轴对产品进行移动,以便对产品进行多次拍照。
(4) 工位顶部避开机台照明灯,以防止外界光对外观检测进行干扰。
(5) 上、下面检测需要布置在两个工位。

总结如下。

(1) 单像素精度无法满足上述要求,但是整体可以实现检测的功能项。
(2) 0.1 mm 的脏污能不能检测?由于 0.1 mm/0.063=1.58,这个值已经小于 2 个像素,所以不能检测 0.1 mm 的脏污。

20.4 硬件配置

硬件配置主要包括相机、光源、镜头的相关配置,见表 20-2。

表 20-2 视觉硬件选型配置表

序号	物料号	硬件类型	品牌	型号/规格	数量	易损件	备注
1		相机	海康威视	MV-CH120-10GC	2	□是 ■否	1 200 W 像素彩色相机
2		镜头	灿锐	XF-MH004	2	□是 ■否	
3		光源	嘉励	JL-DM-352W-C	2	□是 ■否	
4		光源	嘉励	JL-HAR-366W-H16-DB37	2	□是 ■否	
5		面光源延长线	嘉励	JLE-5MGR-Y1-24V	2	□是 ■否	5 m 光源延长线
6		环形光源延长线	嘉励			□是 ■否	
7		光源控制器	嘉励	JL-APS2-14424-2	1	□是 ■否	
8		光源控制器	嘉励	JL-DMS2-200W-16	2	□是 ■否	
9		千兆网线	OPT	OPT-GIGE-10M	2	□是 ■否	10 m 通信线缆,依整体布局可作更改
10		相机电源	三铭	A-6S1-10-D	2	□是 ■否	相机电源
11		网线	秋叶原		3	□是 ■否	用于光源控制器通信

20.5 硬件安装

整体机器视觉的硬件安装示意图如图 20-7～图 20-10 所示。

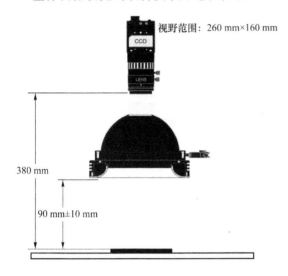

图 20-7 视觉硬件安装示意图主视图（上面）

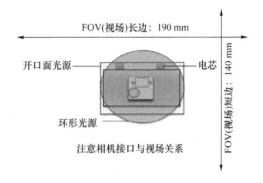

图 20-8 视觉硬件安装示意图俯视图（上面）

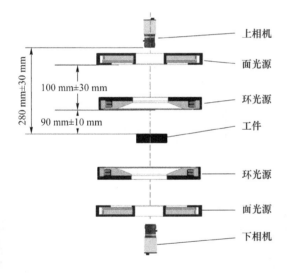

图 20-9 视觉硬件安装示意图主视图（上、下两面）

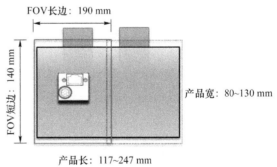

图 20-10 大电芯两次拍照示意
260 mm×160 mm、180 mm×146 mm

20.6 安装图纸

相机安装示意图如图 20-11 所示。

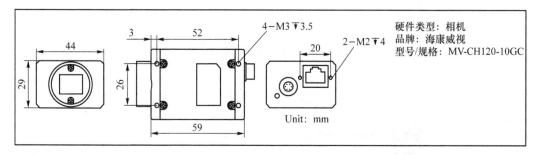

图 20-11 相机安装图纸

镜头安装示意图如图 20-12 所示。

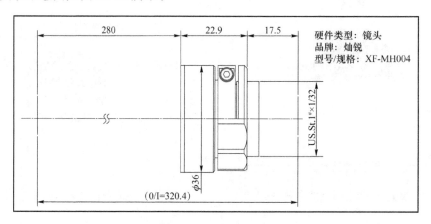

图 20-12 镜头安装图纸

光源安装示意图如图 20-13 所示。

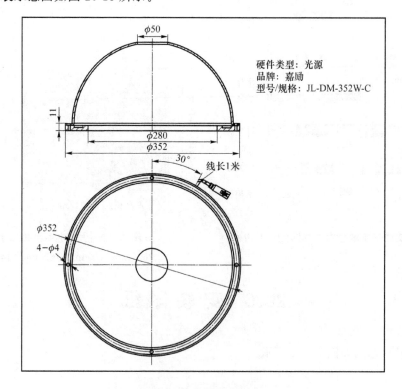

图 20-13 光源安装图纸

一个完整的机器视觉系统由以下组成:判定标准、光源(可见光、X光、激光)、镜头、相机、计算机系统、检测对象和检测结果。本章从机器视觉的原理开始,首先讲解机器视觉可以解决引导、检测、测量、识别四个方面的应用;然后讲解机器视觉系统光源的选择、镜头的原理及选型、相机的介绍及选型,并通过实际案例,动手完成一个项目实战的硬件选型。

机器视觉的选型首先是硬件的选型,然后是软件开发的选型。硬件以相机为核心的选择,主要是巴斯勒、海康、大恒、科视威、大华等为代表的选择;软件以 VisionPro、Halcon、Labview、OpenCV 等为代表的选择。

机器视觉原理就是给设备添加一个由 PLC(或嵌入式系统)出发的信号,通过串口(或网口),通知计算机完成采图工作,然后对图像进行增强、滤波、特征提取后,进行判断 OK 或者 NG 后,将信号通过串口(或网口),回传给 PLC(或嵌入式系统),完成分拣任务。

机器视觉的四类任务:第一,引导贴片机、盲人导航机器人、人脸识别、尺寸检测等待方面的应用。第二,检测包括白瓶检测(检测标签是否存在)、瓶底检测、装箱计数、铝箔表面缺陷检测。第三,测量:钟表轴承测量,身高和体重自动测量。第四,识别,喷码检测,合同审批,智能报销,批阅试卷。其主要应用的领域有汽车、电子、新能源、制药、陶瓷、面板、服装、饮食、农业等。

20.7 专业术语

在相机聚焦完成后,在焦点前后的范围内,都能形成清晰的像,这一距离叫作景深。

光学放大率:像的尺寸/实际物体的尺寸。

CCD 靶面:1/4 inch,1/3 inch,1/2 inch,2/3 inch,1 inch(12.8×9.6)。

镜头接口有两种,一种是 C 口:CCTV,C-Mount,公称直径=1,螺距 32 牙,17.625 mm;另一种是 CS 口:CCD 常用接口,12.5 mm。